JAMAICA

AF767152

PEP

Mathematics

Performance Task

GRADE 4

Practice Papers

hachette LEARNING

The Publishers would like to thank the following for permission to reproduce copyright material.

Photo credits

t = top, b = below, l = left, r = right, c = centre, b/g = background

p.13: © Laura/stock.adobe.com; p.21: © Laura/stock.adobe.com; p.32: © topvectors/stock.adobe.com; p.37: (tl) © pixelrobot/stock.adobe.com, (tl) © Dusty Cline/stock.adobe.com, (cl) jeayesy/stock.adobe.com, (bl) © trekandphoto/stock.adobe.com, (bl) © Olga Kovalenko/stock.adobe.com, (tr) © Bowonpat/stock.adobe.com, (cr) © robynmac/stock.adobe.com, (cr) © Alec/stock.adobe.com, (br) © whatamiii/stock.adobe.com; p.38: © kolonko/stock.adobe.com; p.40: © Lumos sp/stock.adobe.com; p.44: © Prostock-studio/stock.adobe.com; p.58: © Alexander Raths/stock.adobe.com; p.62: © Ms VectorPlus/stock.adobe.com; p.67: © Moving Moment/stock.adobe.com; p.79: © Stewart/stock.adobe.com; p.91: © bablab/stock.adobe.com; p.97: © Neelstudios/Wirestock Creators/stock.adobe.com.

Acknowledgements

Although every effort has been made to ensure that website addresses are correct at time of going to press, Hachette Learning cannot be held responsible for the content of any website mentioned in this book. It is sometimes possible to find a relocated web page by typing in the address of the home page for a website in the URL window of your browser.

Hachette UK's policy is to use papers that are natural, renewable and recyclable products and made from wood grown in well-managed forests and other controlled sources. The logging and manufacturing processes are expected to conform to the environmental regulations of the country of origin.

To order, please visit www.HachetteLearning.com or contact Customer Service at education@hachette.co.uk / +44 (0)1235 827827.

ISBN: 9781036005191

First published in 2026 by Hachette Learning (a trading division of Hodder & Stoughton Limited),
An Hachette UK Company
Carmelite House
50 Victoria Embankment
London EC4Y 0DZ

www.HachetteLearning.com

The authorised representative in the EEA is Hachette Ireland, 8 Castlecourt Centre, Dublin 15, D15 XTP3, Ireland (email: info@hbgi.ie)

Impression number 10 9 8 7 6 5 4 3 2 1

Year 2030 2029 2028 2027 2026

Cover photo © GoodStudio- stock.adobe.com

Illustrations by Oxford Designers and Illustrators Ltd and Hyphen S.A./Konstantina Kyratzidou

Typeset by Hyphen S.A.

Printed and bound in Great Britain by Bell & Bain Ltd, Glasgow

A catalogue record for this title is available from the British Library.

FSC www.fsc.org MIX Paper | Supporting responsible forestry FSC™ C104740

Contents

How to use this book

Think about successful athletes. Athletes learn about their sport, and they practise and train hard to become excellent performers.

You can be a maths athlete who performs well in tasks by learning how to do them and then training yourself to answer different types of questions.

You are going to work through **Practice paper 1** with your teacher. This will help you to:

- Read questions carefully and work out what type of answer to give.
- Understand why it is important to show how you work out your answer.
- Learn how you can get good results even if your answer is not completely correct.
- Understand what the different ratings mean and see how you can improve your rating.
- Reflect on your own work to help you see what you did well and what you can improve in the next task.

Each task in this book has three supporting features.

Before you start …

This section helps you think carefully about questions and what it is you are being asked to do. There are different activities for each task.

Hints

These are given in the task itself to help you think about the questions and how you should answer them. As you progress, there will be fewer hints in the tasks.

Reflection

This section encourages you to think about your own learning and how you can improve on it.

Practice paper 1 also has a section called **How do you know how well you did?**

In this section you will see possible answers to the questions and what rating they were given. The aim of this section is to show you what you need to do to get the best possible results.

Practice paper 1

Always check that you have everything you need before you start.

This task has 3 parts:

- Part 1 has questions 1 and 2.
- Part 2 has question 3.
- Part 3 has questions 4 and 5.

You will need a calculator, a pencil and an eraser.

Answer all the questions in each part in the spaces provided.

Remember to always show your working out.

→ Before you start …

When you do a performance task, it is useful to read through the questions first so you have an idea of what you will need to do.

Start by **skim reading** all the questions. **Circle** or highlight any **instruction words**.

Now, **draw a *** next to questions that look like they have only one short answer.

Next **look at the underlined instructions** in the task.

What do you have to do when you answer these questions? **Share your ideas** with your partner.

At the library

Verona and Sheldon are library monitors. They help in the library after school and at weekends.

Their duties include:

- Checking books in and out of the library.
- Helping people find what they need.
- Returning books to the correct shelves.
- Keeping shelves neat and tidy.
- Organising chairs, turning on computers and making sure people know how to use library technology.
- Ensuring people follow the library rules, including no food or drink in the library.

Part 1

> Look at the map carefully. What does the key tell you?

Going to the library

Question 1

This map shows the area around Verona's house. You are going to use it to work out how she can get to the library.

a Verona is in front of her house. Find her on the map. Put a * next to her house.

b Find two different routes that Verona could take to get to the library. Show both routes clearly on the map. Use different colours or two different styles of lines.

c Approximately how far does Verona walk if she takes the shorter of the two routes? Explain how you worked this out.

> Remember, the map scale tells you how far things are in the real world. You have to measure the map distance and use the scale to find the real distance.

How far will 200 m be on the map? How can that help you find Sheldon's house?

d Sheldon lives approximately 200 m from the library. Which houses could be his? Mark them on the map. Explain how you decided which houses could be Sheldon's house.

Question 2

When you are given information like this, you will need to use it to answer the questions.

The library opening hours are:

Monday	10:00–17:00
Tuesday	09:00–18:00
Wednesday	10:00–17:00
Thursday	10:00–15:00
Friday	09:00–12:00, then 14:00–19:00
Saturday	09:00–15:00
Sunday	**CLOSED**

You need to work with the times and the days given to find the answers to these questions. Draw a number line to help you count hours if you need to.

a Complete each sentence by filling in the correct day of the week:

- The library stays open for the longest time on a __________.
- The library is open for 5 hours on a __________.
- The latest the library stays open is on __________.
- The library closes for a 2-hour lunch break on a __________.

What time does the library open on a Tuesday? Find that information before you answer the question.

b Verona takes 25 minutes to walk to the library. What is the latest she can leave home to be at the library at opening time on a Tuesday?

Part 2

Taking out and returning books

Question 3

Sheldon made this sign for the library desk.

> Read the information carefully. Highlight the important number facts. You will need these to answer the questions.

Junior members (under 15)

You may borrow 1 book at a time.

Books are due to be returned 2 weeks after they are taken out.

Please return the book before or on the due date.

The fine for returning a book late is \$35.00 per week (up to 3 weeks).

If the book is more than 3 weeks overdue or lost, the fine is \$750.

> You know that there are 7 days in a week. You can work out the date one week later by adding 7 to the date you are given.

a Marcia took out a book on 3rd May. What is the latest date she can return it without paying a fine?

> If the book is one week late, it means it was due one week before that. Think about how you can work out that date.

b Janelle returned her book on 23rd May. If the book was 1 week late, when was it due to be returned?

What do you need to know before you can answer this question? Where will you get that information?

c Jerome's book was due back on 1st May. He only returned it on 15th May. What fine would he have to pay?

How late was Marcus' book?

d The librarian collected this data for books returned late in May.

Number of days late	Number of books
1–7 days	12
8–14 days	9
15–21 days	6
Over 21 days	3

i. Which category had the most late books?

ii. How many books were over a week late?

iii. Estimate how much money was collected in fines in May.

Part 3

Using the computers at the library

Question 4

The library loans out tablet computers for use inside the library. If you are under the age of 15, you may only use the tablet in the computer room where there is a library monitor or member of staff. Tablets may be borrowed for up to 3 hours per day.

The computer room has square tables. Each table can fit one chair on each side, like this:

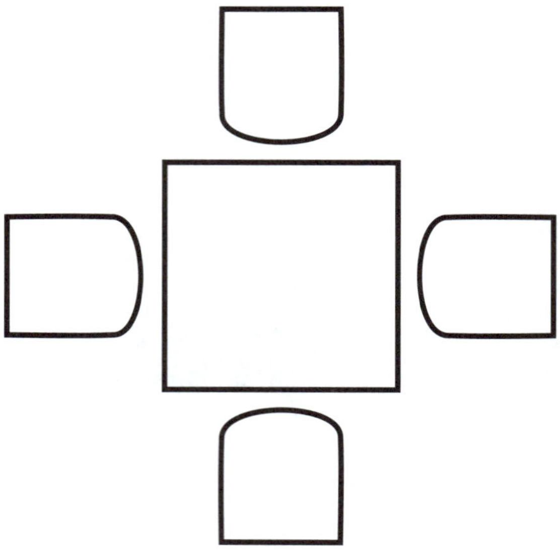

a The librarian arranges the tables so that students can work in groups.

For each of these arrangements of desks (in **A–D**), write how many chairs will be needed.

Draw a dot on the diagram to show each chair and then count them. Remember to write the answers as well.

A

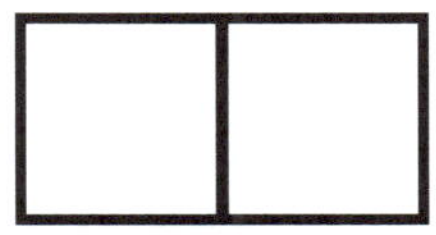

B

C

D

What do you notice about the number of chairs in Group C and D? Can this help you work out the answer here?

Remember you can try out a few ideas before you decide which one is correct.

b Which grouping of six desks will give the greatest number of chairs? Draw your idea here.

Skim read the rules and remember you will need to find the information you need to answer the questions.

Question 5

The library has these rules for using tablets:

DESCRIPTION	FINES/COSTS
Fine for every 15 minutes or fraction thereof that the tablet is overdue. There is no maximum fine.	$50
Lost or damaged power cord	Replacement cost: $2 000
Lost or damaged accessories	Replacement cost: dependent on the cost to replace accessory
Damaged tablet	Repair cost: dependent on the cost to repair damage
Lost or stolen tablet. (*If the tablet is stolen, the borrower is responsible for submitting a police report to the Senior Librarian.*)	Replacement cost: $50 000

How can you work out how long he had the tablet? Could a number line help you count up the time?

a Samuel borrows a tablet at 12:25 p.m. He returns it at 3:40 p.m. without the power cord. What fine will he have to pay? Show how you work this out.

The question asks what you think, but it also asks why you think that. You will need to use the information in the table to work out what will happen.

b Nadia accidentally knocked a tablet off the table and cracked the screen. She reports this to the librarian immediately. What do you think will happen next? Explain why.

c This diagram shows one of the tablets.

How do you work out the perimeter of a rectangle?

Can you draw a model to help you work this out?

The rectangular screen has a perimeter of 210 cm. The length is 60 cm.

a What is the width of the screen?

____

b Mark the length and width on the diagram.

c Explain how you worked it out.

____

____

Practice Paper 1 Self-evaluation

How do you know how well you did?

Now that you have worked through the task, you are going to look at some answers that students gave to see how they were rated. Then you can rate your own answers using this scale.

1 Beginning	2 Developing	3 Proficient	4 Highly proficient
Your answers tell the teacher that you do not understand the work and that you need a lot more support.	Your answers show that you understand some of the work but that you need help in some topics.	Your answers show that you understand most of the work.	Your answers show that you understand all of the work very well.

Question 1

a Verona is in front of her house. Find her on the map. Put a * next to her house.

b Find two different routes that Verona could take to get to the library. Show both routes clearly on the map. Use different colours or two different styles of lines.

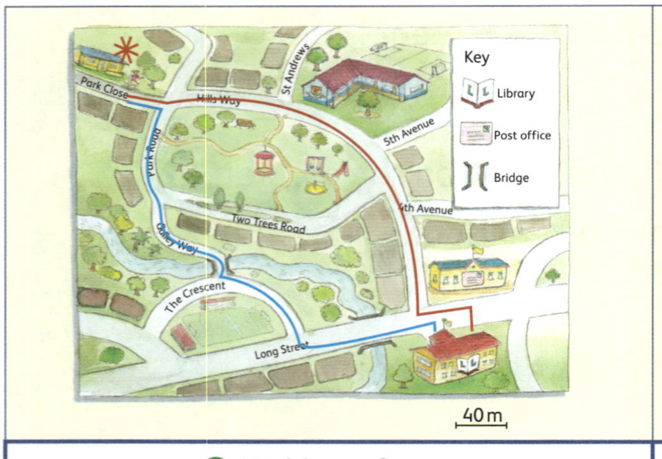

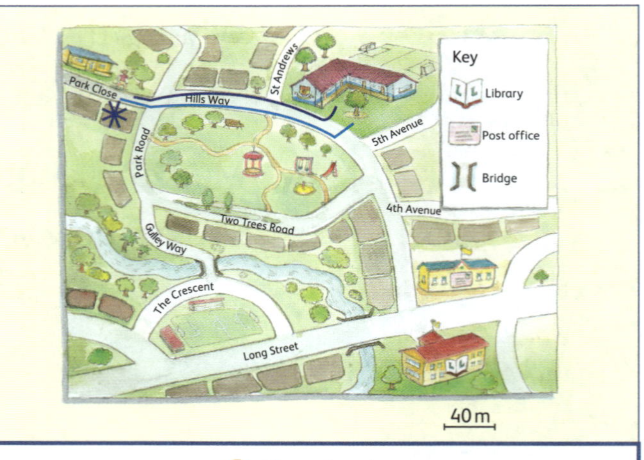

4 Highly proficient	2 Developing
a *The student has clearly marked Verona's house.* ***b*** *They have also used two different colours to show clear routes from her house to the library.*	***a*** *This student has marked a house and it is close to where the girl is standing, but it is across the road, so it is not the correct house.* ***b*** *The student understood the instructions and has shown two clear routes. They did not read the map key properly, so they have not drawn routes to the library.*

c Approximately how far does Verona walk if she takes the shorter of the two routes? <u>Explain how you worked this out.</u>

12 cm

1 Beginning

12 cm is not a sensible answer for real life distances. It shows the student does not understand measurements.

Student has not used the scale and has not explained how they got to the answer.

She walks 650 m.

I measured the route in centimetres and then multiplied my answer by 40 to get metres.

3 Proficient

The student answered the question and gave a reasonable distance. They did not show their working though.

The student did explain clearly what they did to get to the answer.

d Sheldon lives approximately 200 m from the library. Which houses could be his? Mark them on the map. <u>Explain how you decided</u> which houses could be Sheldon's house.

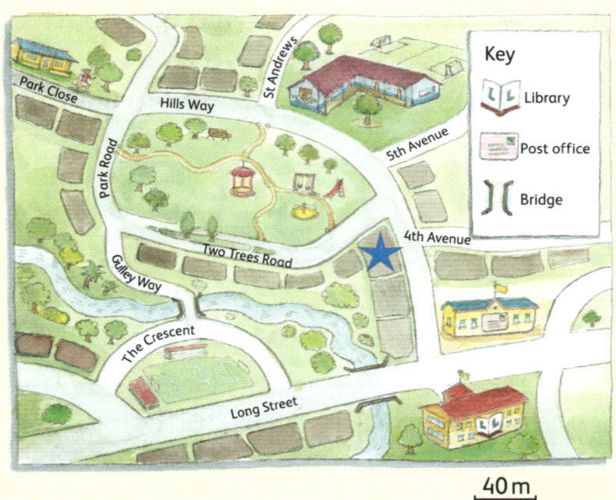

The star shows Sheldon's house.

I worked out that
200 m ÷ 40 m = 5 cm to get the distance on the map. Then I used a ruler to find a house that was 5 cm away from the library.

4 Highly proficient

This student has given a correct answer and explained clearly how they got to the answer using the scale and then measuring on the map.

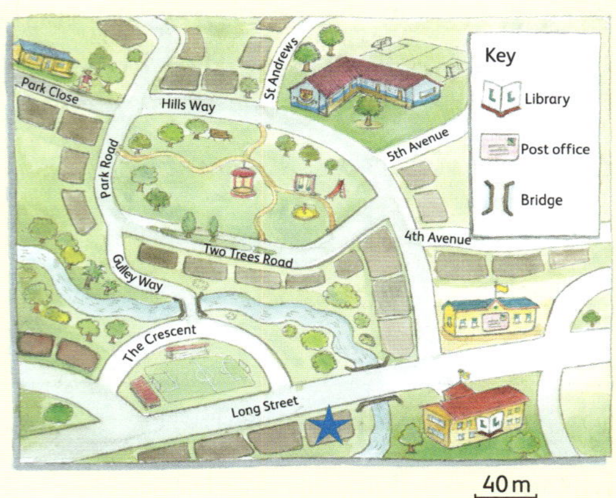

I measured 5 cm from the library to find his house.

3 Proficient

This student has understood the task and used the scale, but has not explained how they knew they had to measure 5 cm. Also, the house they have chosen is only X cm from the library.

Question 2

a Complete each sentence by filling in the correct day of the week.

• The library stays open for the longest time on a Friday. ✗ • The library is open for 5 hours on a Friday. ✗ • The latest the library stays open is on Sunday. ✗ • The library closes for a 2-hour lunch break on a Friday. ✓	• The library stays open for the longest time on a Tuesday. ✓ • The library is open for 5 hours everyday. ✗ • The latest the library stays open is on Saturday. ✗ • The library closes for a 2-hour lunch break on a Friday. ✓
1 Beginning *The student has not managed to find the correct information in the list of opening times.*	2 Developing *The student has tried to make sense of the opening hours but has not managed to get all the answers correct.*

b Verona takes 25 minutes to walk to the library. What is the latest she can leave home to be at the library at opening time on a Tuesday?

09:00 + 25 minutes later is 09:25.	Opening time is 09:00. 25 minutes before that is 08:35. The latest she can leave is 08:35.
1 Beginning *The student has mixed up the latest time with "later" and added 25 minutes instead of subtracting.*	4 Highly proficient *The student has read the correct opening time and calculated the time 25 minutes earlier than that to get the right answer.*

Question 3

a Marcia took out a book on 3rd May. What is the latest date she can return it without paying a fine?

2 weeks later.	She can keep the book for 2 weeks. 2 weeks is 14 days. 3 + 14 = 17 May.
2 Developing *The student knows how long she can keep the book but has not answered the question correctly because they did not give a date.*	4 Highly proficient *The student has the correct answer and has shown how they worked it out.*

b Janelle returned her book on 23rd May. If the book was 1 week late, when was it due to be returned?

30 May.	1 week is 7 days later. It was due on the 16th.
1 Beginning *The student has added on a week instead of working out that the book was due a week earlier.*	3 Proficient *The student understands the question and has the correct answer without the month.*

c Jerome's book was due back on 1st May. He only returned it on 15th May. What fine would he have to pay?

\$35 × 2	That is 2 weeks late, so \$35 × 2 = \$70.
2 Developing *The student understood the question but has not worked out the actual answer.*	4 Highly proficient *The answer is correct and the student has shown how they got to the answer.*

d Marcus took out a book on 1st May and forgot to return it. He only remembered he had it on 31st May and he took it back to the library. What fine would he have to pay? Explain why.

He would have to pay $105.	His fine would be $105. The book was due on 15 May. He took it back 2 weeks and 2 days later, so he would most likely have to pay 3 × $35 = $105.
3 Proficient *The student has the correct answer but has not clearly explained why he would pay $105.*	**4** Highly proficient *The student has given the correct answer and clearly explained why the fine would be $105.*

e The Jamaican Information Services announced that people could return their library books, no matter how overdue they were, during June and July and pay no fines. Why do you think they may have done this?

To make people not pay their fines.	I think they do this to get people to return books they have had for a long time. Sometimes people might not take the book back because they don't have enough money to pay the fines. This helps the library and the people who cannot afford to pay.
1 Beginning *The student has not given an opinion or any sensible reasons.*	**4** Highly proficient *The student has a clear opinion and gives a sensible reason for the Information Services decision.*

Question 4

a The librarian arranges the tables so that students can work in groups. For each of these arrangements of desks (in **A–D**), write how many chairs will be needed.

A 2 B 3 C 4 D 4	A 6 B 8 C 10 D 10
❶ Beginning *The student seems to have given the number of desks and not the number of chairs.*	❷ Developing *The student has most answers correct but seems to think that C and D both have 10 because they both have 4 desks.*

b Which grouping of six desks will give the greatest number of chairs? Draw your idea here.

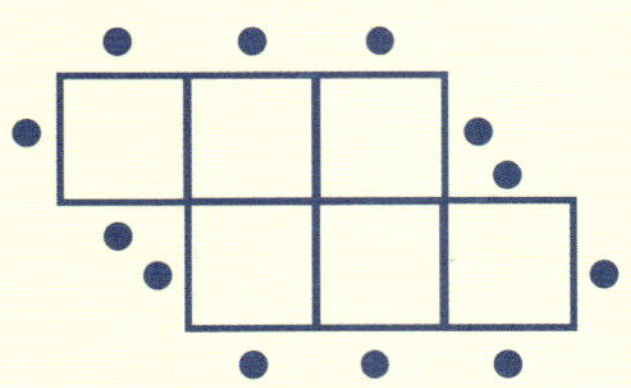 12 chairs are the most.	 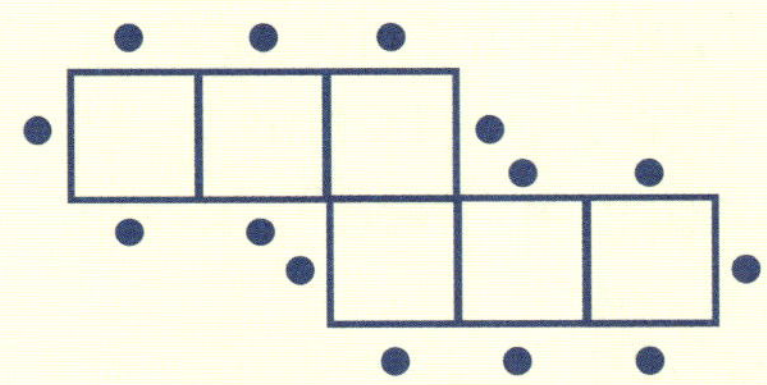 The most chairs you can have with 6 desks is 14. But if you put them like this, you can get 16. 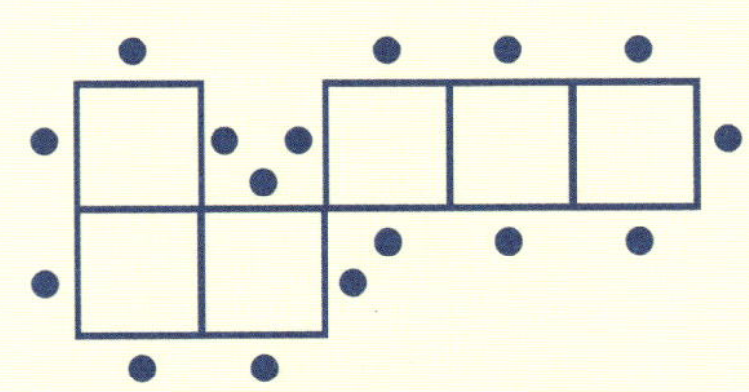
❷ Developing *Even though the answer is not correct, the student shows they understood the question and they have drawn a solution.*	❹ Highly proficient *The student has shown two ways of getting the maximum number of chairs and suggested a different arrangement to get 16 chairs.*

Question 5

a Samuel borrows a tablet at 12:25 p.m. He returns it at 3:40 p.m. without the power cord. What fine will he have to pay?

$50 fine.	$50 fine 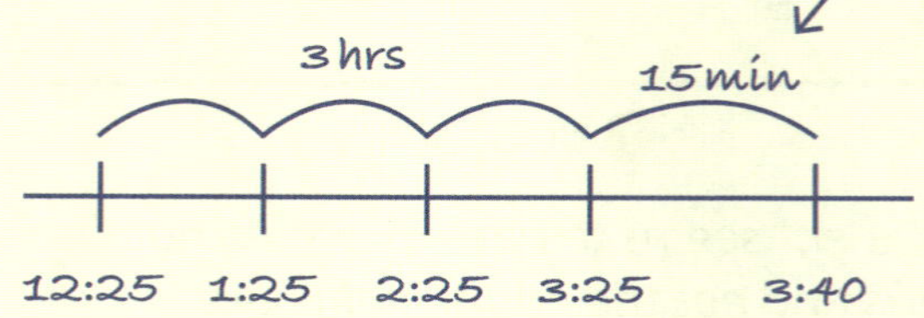 Plus $2 000 for the power cord. His fine will be $2 050.
❶ Beginning *Student has not considered the power cord and they have not shown any working out.*	**❸ Proficient** *Student has drawn a number line to show how they worked with time and they have added the power cord fine.*

b Nadia accidentally knocked a tablet off the table and cracked the screen. She reports this to the librarian immediately. <u>What do you think</u> will happen next? <u>Explain why.</u>

She will have to pay to fix it.	I think the library will take her details and make her sign that she has damaged the tablet. The library will need to find out what the cost will be before Nadia can pay for the damage. Once they know that, Nadia will need to pay.
❷ Developing *The student has said what will happen next but has not said why they think this.*	**❹ Highly proficient** *The student has shown a good understanding of what is likely to happen as a result of Nadia damaging the tablet.*

c Work out how wide and how long the screen is.

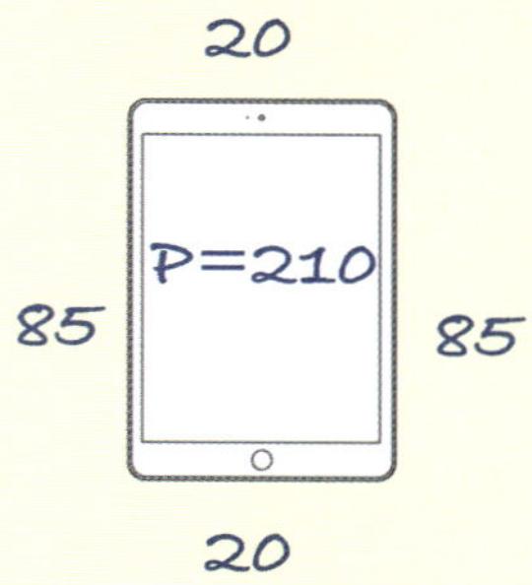

$$\frac{20}{85} \neq \frac{3}{4}$$

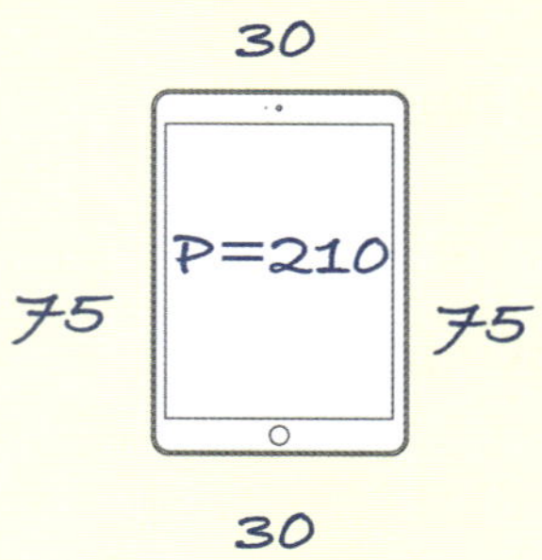

$$\frac{30}{75} \neq \frac{3}{4}$$

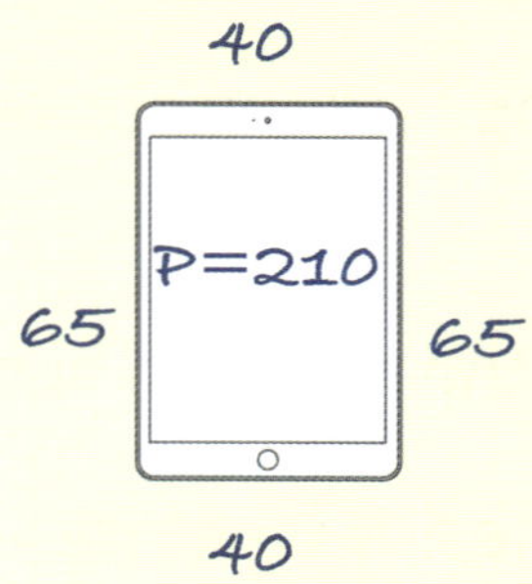

$$\frac{40}{65} = \frac{8}{13}$$

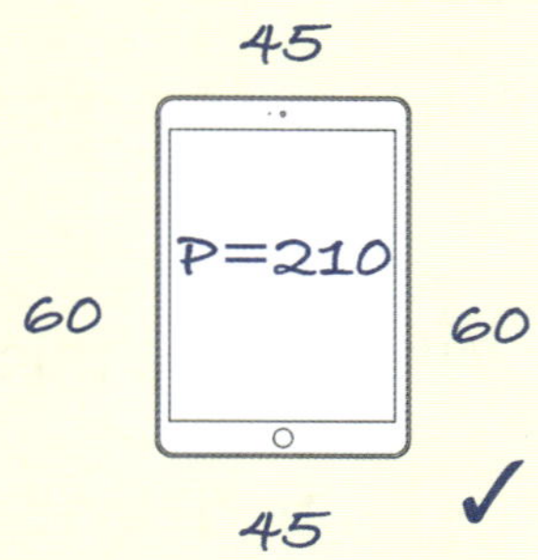

$$\frac{45}{60} = \frac{9}{12} = \frac{3}{4}$$

Perimeter = 2 × (length + width)

So,
length + width = 210 ÷ 2 = 105 cm

The ratio of width to length is $\frac{3}{4}$ or 3 : 4

105 ÷ 7 = 15

3 × 15 = 45

4 × 15 = 60

Check: $\frac{45}{60} = \frac{3}{4}$

45 + 60 = 105 cm

The screen is 45 cm wide and 60 cm long.

2 Developing	4 Highly proficient
The student has used trial and error to work out the correct answer. This is quite a long process, but it shows they understood the question and they got to the right answer.	*The student has used the formula for perimeter and applied what they know about fractions to get the correct answer in a clear and short way.*

Reflection

Go back through your own answers to the questions in this task.

Which question did you find the easiest to answer?

____

- Why do you think the question was easy for you?

  ☐ I had done one like it before
  ☐ I understood what to do
  ☐ I remembered how to do it
  ☐ I used pictures or working out
  ☐ Other: ______________________

Which question did you find the hardest to answer?

____

- Write the question number or what it was about. ______________
- Why do you think the question was hard?

  ☐ I didn't understand it
  ☐ I forgot what to do
  ☐ I made a mistake in my working out
  ☐ Other: ______________________

- What is one thing you can do next time to help yourself?

  ☐ Read more carefully
  ☐ Use the working-out space
  ☐ Ask for help if I need it
  ☐ Take my time
  ☐ Other: ______________________

Practice paper 2

This task has 4 parts:

- Part 1 has question 1.
- Part 2 has question 2.
- Part 3 has questions 3 and 4.
- Part 4 has question 5.

Draw or write your answers to all of the questions in the spaces provided.

Remember to show your working.

Ask yourself: "What am I being asked to do here?"

Jot down your ideas.

→ Before you start ...

Read through all the questions.

Here are some key words that might help you.

Use a formula   Do a calculation
Solve a two-step problem   Draw a graph
Show working on a diagram   Use patterns
Give an opinion   Make decisions   Give reasons

Growing vegetables

The Grade 4 class at St Patrick's Primary school are planning to grow vegetables in the school garden.

The principal has given them a rectangular plot of land to use but the students will measure and mark it out. They will plant the vegetables once they have done that.

Part 1

Marking out the plot

Question 1

The Grade 4 plot is at the back of the school field next to the wall. The measurements of the plot are given on the diagram.

How can you work out the missing measurement?

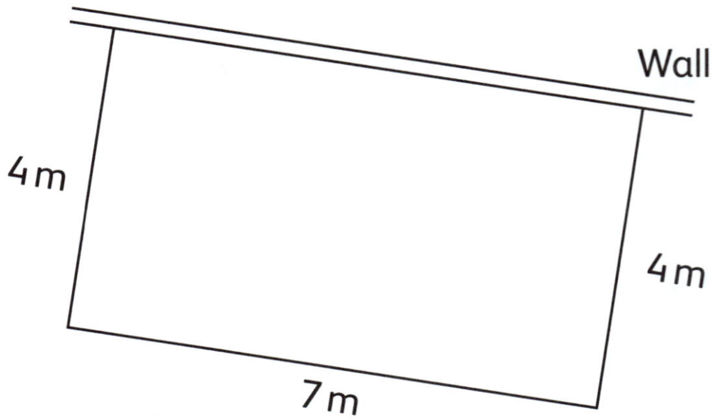

a What is the area of the plot in square metres (m^2)?

The plot is a rectangle. How do you work out the area of the rectangle?

b The students will use string to mark the boundaries of their plot. They do not need to mark the side where the wall is. What length of string will they need?

c There are 4 groups in the class. Each group will get an equal share of the plot. Draw lines on the diagram to show how you can divide the plot into 4 equal parts.

What fraction is each part if there are 4 equal parts?

Part 2

Dividing up the beds

Question 2

The class divide the planting area up into small beds. They put 1 m long wooden planks around the beds like this:

The numbers tell you how many planks they use for each bed.

Toniqua draws this diagram to show how many planks her group needs for 4 beds.

4	3	3	3

a Why do they need 4 planks for the first bed but only 3 for the others?

b How many planks will they need to make 7 beds in a row?

c Another group makes 4 beds like this:

How many planks will they need for 8 beds? Show how you work this out.

A drawing might help you work this out.

Part 3

Organising the plants

Question 3

Jameel's group has 3 beds in a row like this:

Think about how many choices you have for the first, second, and third positions. Try starting with a different letter each time and use your working-out space so you don't repeat combinations.

They are planting:

- **O** for okra
- **P** for peppers
- **T** for tomatoes

They want to try out different ways to plant the 3 vegetables, using one plant in each bed.

Write as many different ways as you can to arrange the plants. Use the letters O, P, and T.

Here's one way:

T – P – O

Now try to find the others.

b How many different arrangements did you find?

Write the number: ______

c Do you think you found *all* the possible ways?

How can you be sure? (Give a reason.)

______

Question 4

Use dots or small crosses to represent one plant.

a Sondra's group has 24 pigeon pea plants. They want to plant them in rows, with the same number of plants in each row.
Draw two different ways that they can do this:

You will need to think about equivalent fractions to do this.

b This plot is going to be planted up with dasheen, scotch bonnet pepper, okra, pigeon peas and cabbage. Show how you can divide the plot into the fractions given. Write the names of the plants in each part. Work out what fraction of the plot will be planted with cabbage.

$\frac{1}{3}$ dasheen

$\frac{1}{4}$ scotch bonnet

$\frac{1}{8}$ okra

$\frac{1}{8}$ pigeon pea

$\frac{\square}{\square}$ cabbage

Part 4

See how they grow

Question 5

The class do an experiment to measure what happens when you give plants a spoonful of organic plant food each week.

They record their results in a table.

Week	Plant A – no added plant food	Plant B – 1 spoonful of plant food per week
1	$3\frac{1}{2}$ cm	4 cm
2		
3		
4		
5		
6		

Would it help if you filled in the rest of the table?

The students discover that the plants grow at different rates. The plant with no food grows $\frac{1}{2}$ cm per week. The plant that gets food grows $1\frac{1}{4}$ cm per week.

a How tall will each plant be in week six?

Plant A ____________ Plant B ____________

b Find a method of calculating how tall Plant A will be after 10 weeks if it continues to grow like this.
Show how you do this.

Remember that 1 week = 7 days.

c It takes 21 days for the Scotch Bonnet plants to start sprouting. After that, it takes about 60 days for the plants to produce peppers. If the students plant seeds at the start of the term, estimate how many weeks they will need to wait to harvest the peppers.

Reflection

Looking closely at your mistakes can help you learn and improve your work.

Think about any questions or parts of questions that you got wrong. Which ones did you do well?

- Did you understand what the question was asking? If not, what can you do next time to help you understand better?

- Did you make any mistakes in your working-out? If yes, what can you do next time to check your work?

Practice paper 3

This task has 4 parts:

- Part 1 has questions 1 and 2.
- Part 2 has question 3.
- Part 3 has questions 4 and 5.

Answer all the questions.

Write or draw your answers in the spaces provided.

Remember to show your working.

→ Before you start …

Skim through each question.

- Circle any number information in the question. Remember number information might be given in figures (5, $\frac{1}{2}$ or 1.2) or in words (four, three-quarters, nine kilometres).
- Underline key words in the question. These are clues that tell you what you need to do.
- Highlight any important information that is given in the question. You might need to use this to find the answer.

Sports and games

Ms Florian is organising a youth games event. She invites youth clubs from around the country to attend.

You are going to help her and the youth clubs solve some problems involving organising the event and spending money.

Part 1

Making teams

Question 1

Five community youth groups will take part in a sporting competition.

Each group lets Ms Florian, the organiser, know how many students will be taking part.

Falmouth Youth	Buff Bay YC	Mandeville Youth	Cambridge YC	Kingston Youth
26	28	31	32	35

You are being asked what you think here.

a Ms Florian says, "That is more or less 5 groups of 30, so I estimate we will have about 150 students taking part."
How do you think Ms Florian got to that estimate?

b Work out the exact number of students that will take part.

c Was Ms Florian's estimate reasonable? Explain why or why not.

Question 2

During the competition, all the students will take part in four different sports. The sports and the number of students per team are:

- mini soccer 6 per team
- basketball 5 per team
- cross country 3 per team
- volleyball 4 per team.

a Which sport will have the most teams? How do you know this?

b How many volleyball teams can Ms Florian make?

Think about what calculation you can use to work this out.

c Basketball teams have 5 players. If there are 152 students, how many full teams can Ms Florian make and how many students will be left over?

Part 2

Marking time

Question 3

The number line represents the cross-country race.

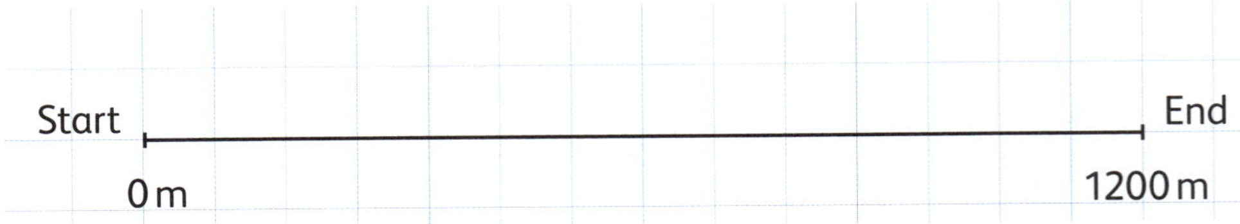

a What is the race distance in kilometres?

Use the grid to help you.

b There are water points at the quarter, halfway and three-quarter marks. Add these to the number line and label them.

c How far is the three-quarter mark from the start of the race? ________

d Each runner in the team covers the same distance. Show how far each team member will run on the number line.

The times on the number lines show how long each of the three runners took to finish their $\frac{1}{3}$ of the race.

e The time taken by each runner in the top two teams is shown here.

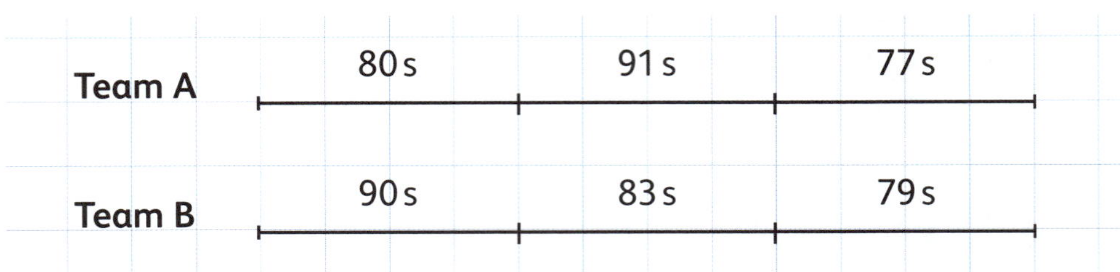

Which team won the race? How did you decide?

f The basketball games are 40 minutes long. If a game starts at 2:25 p.m., when will it be over?

g Each game lasts 35 minutes. Draw a number line to show what time it started if it ended at 2 p.m.

h The runners want to plan the race route for next year. They have three options for the route:

- A straight path
- A loop around the park
- A zigzag through the trees.

i. Draw simple shapes or diagrams to show what each route would look like.

ii. Which route would you choose and why?

Part 3

Wining combinations

Question 4

Each youth club is given a $55 000 voucher to spend on sporting equipment.

a Falmouth decide to spend $\frac{1}{2}$ of the voucher on soccer equipment and $\frac{1}{4}$ each on basketball and tennis equipment.
How much will the club spend on tennis equipment?

b The Buff Bay Youth Club decides to spend $\frac{1}{5}$ of their voucher on sports clothing. How much is this?

c Mandeville spend $18 299 dollars on volleyball equipment. How much do they have left over?

d Cambridge Youth Club spends $\frac{7}{10}$ of the money on new basketball equipment. How much is this in dollars?

To find $\frac{7}{10}$ you can divide by 10 to find $\frac{1}{10}$ and then multiply that amount by 7.

Question 5

Kingston Youth Club gets these prices from a local sporting goods store.

JERRY's SPORTING GEAR PRICE LIST

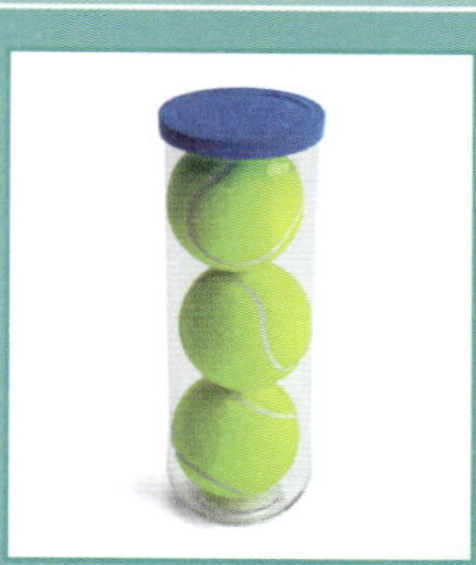

Tennis balls – can of 3 balls
- Match quality = $1 840.16
- Practice balls = $650.97

Tennis racquet
- $17 396.79
- $7 365.28

Football / soccer ball
- $4 113.79

Cricket ball
- $3 352.68

Cricket bat
- $10 014.75

Netball
- $4 810.96

Baseball x 6
- $3 281.58

Shuttlecocks (12 pack)
- $3 812.68

Badminton racquet
- $8 787.98

* Prices include 15% CGT.

The club ordered the following goods:

- 2 netballs
- 3 packs of practice tennis balls
- 12 baseballs
- 4 soccer balls
- 4 badminton racquets
- 1 pack of shuttlecocks
- 1 cricket bat.

Write the prices and the total costs on the list to keep track before you do the invoice.

a Complete this invoice to show what they will pay.

INVOICE

JERRY's SPORTING GEAR
Midway Mall
Kingston
Phone: 0098765
Email: jerry@jsg.com

DATE

INVOICE NO: PQ213

INVOICE TO:
Kingston Youth Club
Half Way Tree

PAYMENT TERMS: Due on receipt

QTY	DESCRIPTION	UNIT PRICE	LINE TOTAL
		$	$
		$	$
		$	$
		$	$
		$	$
		$	$
		$	$
		SUBTOTAL	
		10% DISCOUNT FOR CLUBS	
		TOTAL	

b The club will use the \$55 000 voucher to pay part of the invoice. They will have a raffle to raise the outstanding balance. The raffle tickets cost \$60 each. How many will they need to sell to reach their goal?

Reflection

You can get good marks in this task by clearly showing your working and methods even if you get the final answer incorrect.

Look back at your answers to the task.

- Did you show your working? ______________________
- Is your working clearly laid out and easy for someone else to understand?
  ____
- Does your working show that you understood the question and show that you know how to answer it? ______________________

Practice paper 4

> Read all the information in each part carefully. You will need to use the information to answer the questions.

This task has 3 parts:

- Part 1 has questions 1 and 2.
- Part 2 has questions 3 and 4.
- Part 3 has questions 5 and 6.

You must answer <u>all</u> the questions in each part.

→ Before you start …

Complete these three multiple choice questions by circling (A), (B), (C) or (D). This will remind you about the types of questions you might get in a performance task and how you can think about and answer each type.

1 If the length of rope A is 37 cm, what is the best estimate for the length of rope B?

Rope A

Rope B

To answer this question, I would:

(A) Double the length of rope A to find the length of B.

(B) Use a ruler to measure rope B in centimetres.

(C) Compare the ropes to work out how much longer B is.

(D) Use the scale to find the length of the ropes.

2 Harry collected 23 seashells on the beach. Gina collected 38. How many more seashells does Gina have than Harry?

To answer this question, I would:

(A) Write a number sentence to add the two amounts.

(B) Write a number sentence to find the difference between the amounts.

(C) Draw a number line to count on and find the difference.

(D) Draw a number line to count on and find the total.

3 Tiana and her friends play a game. They take turns on the swing and time each other. The table shows how long each child spent on the swing.

Tiana	Bella	Jan	Kierra	Ulanta
5 minutes	7 minutes	10 minutes	9 minutes	8 minutes

What is the mean number of minutes spent on the swing?

To answer this question, I would:

(A) Put the times in order and find the middle time.

(B) Add all of the amounts together to find the total.

(C) Add the amounts and divide the total by five.

(D) Compare the times to find the most common time.

Screens and things

Mrs Bernard is a teacher of class 4B. She gives her students a project to do.

Help her and the class make sense of the tasks they do and help them make sense of the answers.

Part 1

Internet access

Question 1

Jourdain asks his classmates in 4B to tell him where they access the internet on most days. He collects all their responses and draws a bar chart to show what they say.

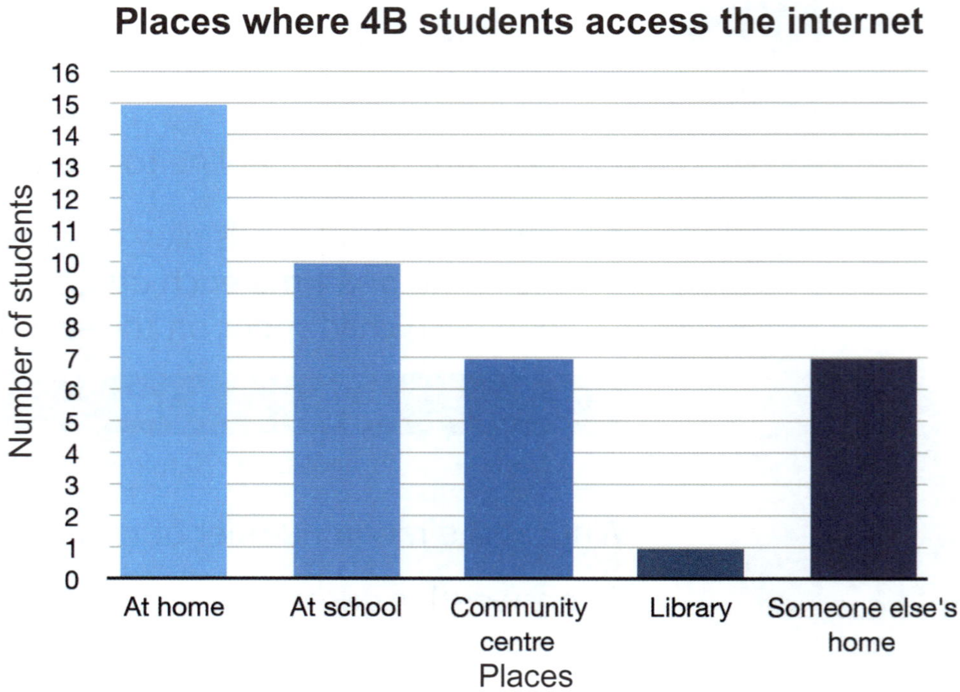

Remember to read the labels and use the scale.

a Where do most of the students access the internet?

____

b How many students access the internet at someone else's home? ____________________

c How many students are in class 4B? Show how you work this out.

What would this fraction be in its simplest form?

d What fraction of the class access the internet at school? Draw a diagram to show this fraction.

Question 2

Do all the bars need to change? Think before you draw.

Five of the students who say they access the internet at home, change their answer. Three of these five students now say that they access the internet at the library and the remaining students say they access the internet at a community centre. Help Jourdain redraw his bar chart to show these changes.

Places where 4B students access the internet

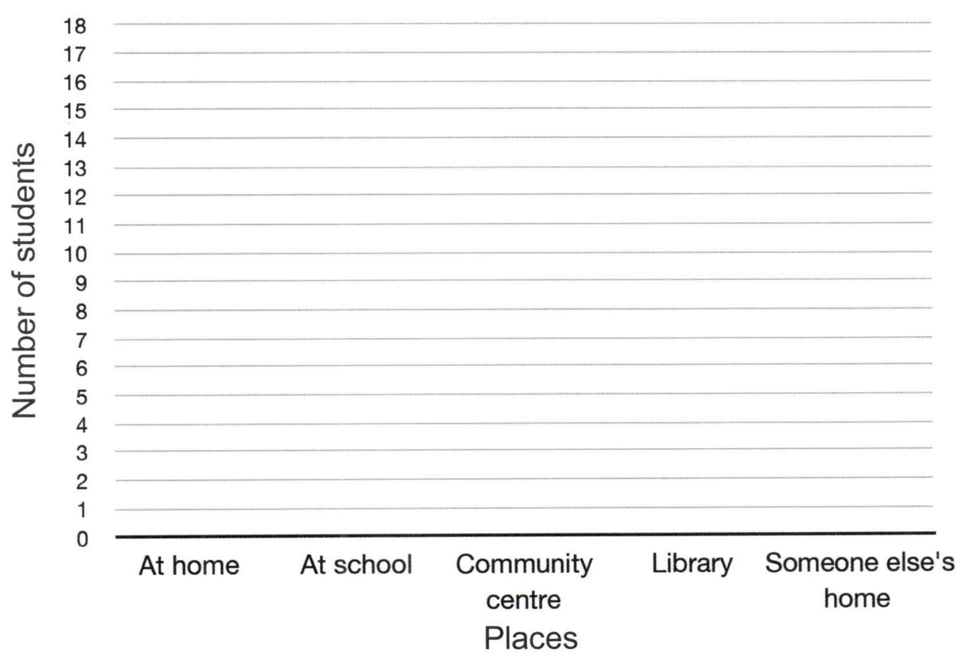

Part 2

Bright Stars

With permission from parents, Mrs Bernard creates a YouTube channel for her class. The name of the channel is Bright Stars. The students are very excited. Together, they create short videos which they upload to the channel.

Here is the data on the Bright Stars YouTube channel after one month.

Channel analytics

Overview | Content | Audience | Research

Your channel got 2 555 views in the last 28 days

Views	Watch time (hours)	Subscribers
2.6K	300	+29
About the same as usual	*more than usual*	*21% more than previous 28 days*

Overview | Content | Audience | Research

Age and gender
Views - Last 28 days

Female	62.2%
Male	40%

In one month, 60% of those who watch Bright Stars are female

In one month, 40% of those who watch Bright Stars are male

Mrs Bernard at the start of the class video on the "Commutative property"

13–17 years	0%	45–54 years	27.9%
18–24 years	0%	55–64 years	20.0%
25–34 years	13.3%	65+ years	10.1%
35–44 years	28.6%		

Each time a person watches a video on YouTube, it is called a "view". Ten views on a video mean that the video has been watched ten times.

Make sure you understand what all the information on the screenshot means.

Question 3

a How many views did Bright Stars channel receive in one month? ______

b Estimate the number of times Bright Stars was watched by males.

c Explain how you know you have a very good estimate.

d How many subscribers did Bright Stars gain in one month? How does this compare with the previous month?

______

Question 4

The picture below shows what the screen looked like when a video from the channel was paused.

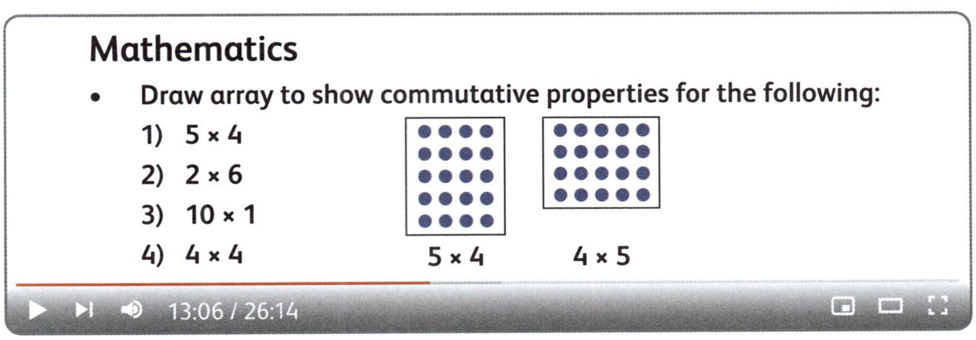

a The video has been paused at 13 minutes and 6 seconds (13:06). Do you think the full video is closer to 20, 25 or 30 minutes?

b Explain how you estimated the total length of the video.

c The video already has 213 views. How many more views are needed for the video to reach 500 views?

The table shows how each video performed on Bright Stars:

Video	Views
Commutative property	213
Fraction magic	500
That does not ADD up	614
Measure me	750
Calendar numbers	?

Think about the total number of views to help you.

d How many views did the video titled "Calendar numbers" receive? How can you check that you are correct?

e What is the mean number of views for the videos on Bright Stars?

Part 3

Device tally and screen time

Mrs Bernard asks her class to write down the devices they own. If a student does not own a device, they write "none".

Here are the responses of the 4B students.

cell phone	laptop	cell phone	tablet
laptop	cell phone	tablet	tablet
tablet	tablet	cell phone	tablet
cell phone	laptop	laptop	none
cell phone	none	cell phone	laptop
laptop	laptop	laptop	laptop
cell phone	laptop	cell phone	none
tablet	tablet	cell phone	laptop
tablet	none	cell phone	laptop
none	tablet	laptop	cell phone

Question 5

a Create a tally chart to organise this data. Ensure that your tally chart has a title.

b Fill in the missing word or number in the spaces below.

i. Most students own a ________.

ii. ________ students do not own a device.

iii. One-quarter of the class owns a ________.

iv. The total number of students in 4B is ________.

You can find all the answers in your tally table.

Students from another class, 4A, also told Jourdain which devices they own.

- All the students in 4A own a device.

- Each student in 4A either owns a tablet or a cell phone.
- In 4A, two times as many students own tablets than in 4B.
- The number of students who own cell phones in 4A is half the number of students who own cell phones in 4B.

Explain as clearly as you can.

c How many more students are there in 4B than in 4A? Show how you know.

Question 6

Mrs Bernard shares this diagram of how students from another class, 4C, spend their free-time hours each week.

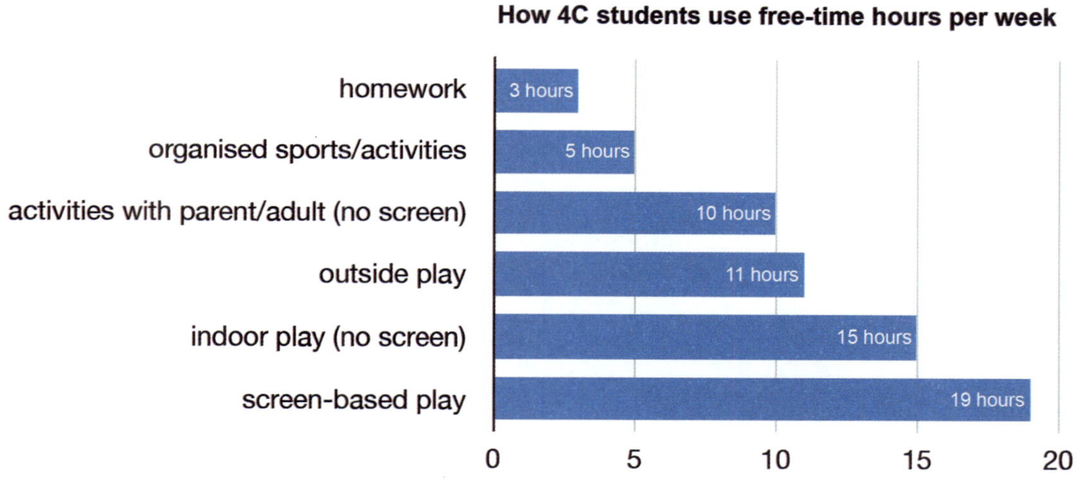

Check your work. Did you write three sentences?

a Write three sentences summarising how the students spend their free time.

Remember there is more than one way to do this.

Marcia's parents feel that she spends too much time on screen and playing, and too little time doing homework.

They decide to draw up a timetable to help her manage her time each day.

b Help them by completing the timetable for Marcia.

<table>
<tr><th colspan="8">Marcia's after-school and weekend timetable</th></tr>
<tr><th>Time</th><th>Monday</th><th>Tuesday</th><th>Wednesday</th><th>Thursday</th><th>Friday</th><th>Saturday</th><th>Sunday</th></tr>
<tr><td>8–9 a.m.</td><td colspan="5" rowspan="6">At school</td><td></td><td></td></tr>
<tr><td>9–10 a.m.</td><td></td><td></td></tr>
<tr><td>10–11 a.m.</td><td></td><td rowspan="3">Marcia and her family go to church</td></tr>
<tr><td>11–12 p.m.</td><td></td></tr>
<tr><td>12–1 p.m.</td><td></td></tr>
<tr><td>1–2 p.m.</td><td></td><td></td></tr>
<tr><td>2–3 p.m.</td><td></td><td></td><td></td><td></td><td></td><td></td><td></td></tr>
<tr><td>3–4 p.m.</td><td></td><td></td><td></td><td></td><td></td><td rowspan="2">Visit to Grandma's house</td><td></td></tr>
<tr><td>4–5 p.m.</td><td></td><td></td><td></td><td></td><td></td><td></td></tr>
<tr><td>5–6 p.m.</td><td></td><td></td><td></td><td></td><td></td><td></td><td></td></tr>
<tr><td>6–7 p.m.</td><td></td><td></td><td></td><td></td><td></td><td></td><td></td></tr>
<tr><td>7–8 p.m.</td><td></td><td></td><td></td><td></td><td>Read a bed time story</td><td></td><td></td></tr>
<tr><td>8:00 p.m.</td><td>BED TIME</td><td>BED TIME</td><td>BED TIME</td><td>BED TIME</td><td>BED TIME</td><td>BED TIME</td><td>BED TIME</td></tr>
</table>

Here are some things to remember:

i. She has dance practice on Tuesday and Thursday.
ii. Dance ends at 4:00 p.m. on Tuesday and 5:00 p.m. on Thursday.
iii. She needs to do 1 hour of homework each day from Monday to Thursday.
iv. Marcia can have screen time (tablet or TV) after she finishes her homework.
v. On Friday, she doesn't have homework and can play outside or watch TV.

Give your own opinion but make sure you give a reason for your answer.

c Fill in the timetable for Marcia from Monday to Friday. Use the boxes to write:

- Homework
- Dance
- Screen Time
- Outdoor Play

Day	3:00–4:00	4:00–5:00	5:00–6:00
Monday			
Tuesday			
Wednesday			
Thursday			
Friday			

d On which day does Marcia finish her homework the earliest?

e How many hours does she spend on screen time from Monday to Friday?

f Do you think Marcia has enough time to relax and play? Why or why not?

Reflection

There are six questions in this task.

Use this scale to rank the questions in order from the one you found easiest to the one you found most challenging.

Easiest					Most challenging

Complete these sentences.

I found question __________ easiest because ______________________________.

I think question __________ was challenging for me because

____.

Practice paper 5

Remember to show all your working and to give reasons for answers if the question asks for them.

When you do a task or test, you can answer the questions in any order.

It is better to do the questions that are easy for you first. That way you have more time to spend on the more challenging parts of the task.

This task has 3 parts:

- Part 1 has questions 1, 2 and 3.
- Part 2 has question 4.
- Part 3 has question 5.

Answer all of the questions in the spaces provided.

→ Before you start …

Read through all the questions.

Here are some key words that might help you.

This task has five questions.

Skim through the questions and decide on an order to answer them.

Write your order here:

First, I will do question ________.

Then, I will do question ________.

After that, I will try question ________.

Next, I will finish question ________.

Lastly, I will do question ________.

Give one reason why you chose this order. Write it here.

____

____

____

____

Finding out about fish

The Ministry of Agriculture has a campaign to encourage Jamaicans to farm with fish. The Grade 4 students decide to investigate what is involved in farming tilapia.

Part 1

Make sure you understand what information the table is giving you. Read the headings carefully.

Where are the fish farms?

The students find this information about fish farming in different parishes.

Parish	Metric tonnes of tilapia produced	Acres in production	Number of farmers registered
St Catherine	290	500	10
St Elizabeth	57	132	3
St Ann	0	1	1
Portland	$\frac{1}{3}$	$\frac{1}{4}$	2
St Mary	0	$\frac{1}{4}$	86
Clarendon	108	78	3
St Thomas	$\frac{1}{3}$	$2\frac{1}{2}$	1
Westmoreland	$\frac{1}{50}$	$4\frac{1}{2}$	4
Hanover	0	$6\frac{1}{2}$	5

Question 1

Write true or false next to each statement. Support your answer with information from the table.

Check your work. Did you give information from the table to support each answer?

a There are no fish farms in Hanover.

____

b St Catherine produces the most tilapia.

____

c There is only one fish farmer in St Thomas.

____

d Westmoreland has a greater area in production than Hanover.

____

e St Mary has more registered farmers than St Elizabeth.

____

Question 2

Use information from the table to answer these questions.

a How many registered fish farmers are there in Jamaica?

b How many more acres are under production in St Catherine than in St Elizabeth?

How can you work out the difference in the number of acres?

c What is the mean number of acres per farmer in Clarendon?

d 1 acre is an area of 4 046 m^2. Estimate the size of the fish farms in Portland and St Mary.

e One metric ton is 1 000 kilograms. How many kilograms of tilapia were produced in Westmoreland?

Question 3

A fish farmer in St Catherine has fifteen 1-acre ponds and each pond has approximately 15 000 tilapia.

a How many acres does the fish farm have in production?

b What percentage is this of St Catherine's total production area?

You need to use the table again to answer this.

c Write the calculation you could do to work out the total number of tilapia on this farm.

d Fill in the missing numbers in this paragraph.

The average weight of one adult tilapia is 330 grams. So, three fish weigh about ________ kg. Each pond on the farm has 15 000 fish. That is about ________ kg of fish in each pond.

Part 2

When you are given information like this, it is helpful to underline or highlight all the numbers or measurements.

Stocking the ponds

Read the information carefully before you answer the questions.

Article Talk Read Edit Tools

Baby fish are called fry. One fry costs \$4.

When you start to stock a fish farm, you need two fry per m^2 of pond. Each fry should be at least 5 cm long.

Once the fish have grown, you can move them to a nursery pond. You need four female fish for every one male in the nursery pond.

Question 4

This is the plan of some of the ponds on Trey's fish farm.

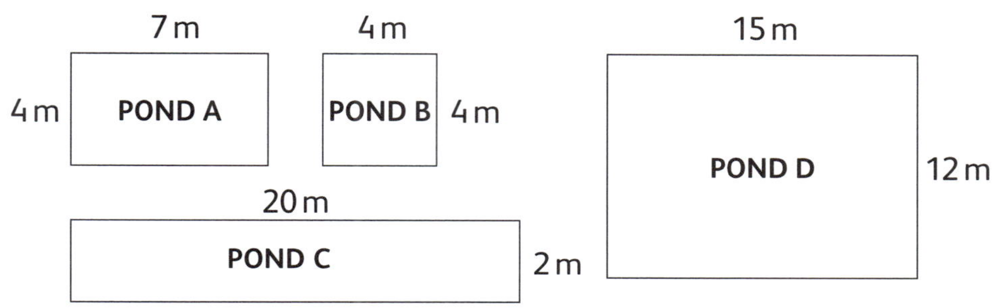

a Calculate the cost of stocking Pond A, B and C with fry.

b Pond D is the nursery pond. Trey wants to stock this with 2 500 fish. How many of these fish should be females?

Part 3

Selling the fish

Question 5

All of the tilapia farmed in Jamaica is sold locally. Most farmers sell to vendors who take the fish to market. Farmers are paid $960 per kilogram by vendors. Markets and other stores sell the fish for $1 430 per kilogram.

If you are not told what type of diagram to draw, you have to think about what type will show the information clearly.

a Draw a diagram to compare the price paid to farmers with the price of fish sold in stores.

b Give one reason for the price difference.

____

c What would a person pay for the following fish at the supermarket if the price was $1 430 per kilogram?

d Hotels and restaurants prefer to buy fish that are between 230–340 grams.

Why do you think they want this size of fish? Give one reason.

Give your own ideas but remember to say why you think this.

Reflection

Once you have checked your answers to the task, complete these statements about your work.

- I now know ________.
- I need to know more about ________.
- These things went well ________.
- I could do better if I ________.
- In my next task I will pay attention to ________.

Practice paper 6

This task has 4 parts:

- Part 1 has question 1.
- Part 2 has questions 2 and 3.
- Part 3 has question 4.
- Part 4 has question 5.

→ Before you start ...

In this task, you are given information at the start of each question.

Find the information in each question.

Highlight all the numbers in the information.

Answer these questions.

- How do you think you will use the information in each question?
- Why is it important to read the information carefully before you answer the questions?

Making pancakes and muffins

The Grade 4 students are very excited because they are taking part in a community project to learn how to make different sweet items which they will sell at the school tuck shop.

You are going to help them understand instructions and make decisions about what they will need and what they should do.

Part 1

Favourite things to make

Question 1

It is easier to work with totals than tallies, so you can complete the table before you answer the questions.

Keshia and Rayshawn did a survey to find out what students in their class would like to make. They made a tally table and started to draw a graph.

What should we make?	Tallies	Total
Grater cake	𝍸	
Toto	𝍸 II	
Pancakes	𝍸 𝍸	
Bulla	𝍸 I	
Muffins	𝍸 III	

	What should we make?
Grater cake	
Toto	
Pancakes	
Bulla	● ● ●
Muffins	● ● ● ●

Key

● = 2 students

a Use the information in the tally table to complete the graph.

b Answer these questions about the data.

What two things are most popular with the students?

________________ and ________________.

How many students would like to learn to make toto?

____

How can you work out how many students took part in this survey?

Part 2

Making pancakes

Question 2

The class decide to make mini pancakes first. They use this recipe.

Mini pancakes

(Makes 12 small pancakes.)

1¼ cup of flour
2½ teaspoons of baking powder
2 tablespoons of sugar
1 egg
¾ cup of milk

Mix the flour, baking powder and sugar in a bowl. Make a hollow in the centre and add the egg and milk. Mix well until the batter is almost lump free.

Melt a little butter in a frying pan.

Put 1 tablespoon of mixture per pancake into the warm pan.

Cook until bubbles appear on the surface, then flip the pancake and cook the other side until it is golden (usually around 1–1½ minutes).

Repeat with the remaining mixture.

Skim read the information before you answer the questions.

The first batch of pancakes does not turn out well.

- Keisha's pancake mixture is very runny.
- Rayshawn's pancakes are too sweet.
- Mira's pancakes are full of bubbles.

a Which ingredient do you think each student measured incorrectly to get those results? Give a reason for your answers.

Keisha ______________________________

Rayshawn ______________________________

Mira ______________________________

b Tareeq's pancakes were very lumpy. What mistake do you think he made?

____

c What advice would you give the students to help them make sure their next batch of pancakes turns out well?

____

How many times does 12 go into 36? How does this help you?

d The recipe the students have is enough for 12 pancakes. How much of each ingredient will they need if they want to make 36 pancakes?

__________ cups of flour

__________ teaspoons of baking powder

__________ tablespoons of sugar

__________ eggs

__________ cups of milk

e If you need $\frac{3}{4}$ of a cup of milk for 12 pancakes, how many can you make if you have $1\frac{1}{4}$ cups of milk? Show how you decide.

Show how you get to your answer. Remember you will get credit for your method even if you get the wrong answer.

f If each person is going to eat 3 mini pancakes, how many batches of the 12-pancake mixture will you need to make enough for 16 people?

Question 3

Rayshawn has a frying pan that can cook 4 mini pancakes at a time. He decides that it takes too long to cook 4 pancakes at a time, so he pours enough mixture into the pan to cook 1 large pancake that he cuts into smaller pieces.

Think about what this means.

Rayshawn discovers that he only has enough mixture to cook 2 large pancakes.

a How many mini pancakes is one large pancake equal to?

b What fraction of the large pancake should Rayshawn give to someone who wants 4 mini pancakes?

Part 3

Making muffins

Question 4

The class decide to make muffins next. The instructions on the recipe tell them to preheat the oven to 375 °F.

The oven they are using has temperatures marked in degrees Celsius.

Start at 275 and look at how much is added each time. Write the missing numbers.

a Fill in the Fahrenheit numbers on the number line.

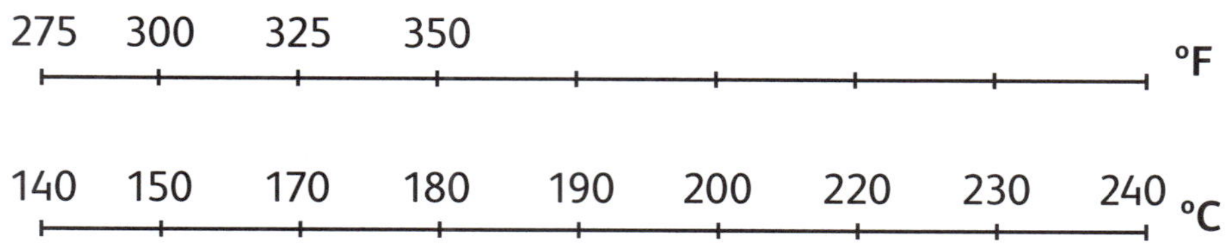

b To what temperature should they preheat their oven?

____

c The oven has a fan. You need to make it 20 degrees cooler than the normal temperature. Write the new temperature here. __________

d The recipe also contains this information:

Prep Time:	**Cook Time:**
10 minutes	25 minutes

How long will it take to make the muffins?

When you work out times, you can use hops on a number line to show your working out.

If the students put the muffins in the oven at 11:15 a.m., when will they be cooked?

e The students have a plate of muffins. The ones in the pink cups are berry muffins and the ones in the yellow cups are banana muffins.

Read what each student says.

Rayshawn	Keisha	Tareeq
There are 2 berry muffins for each banana muffin.	Three-quarters of the muffins are berry muffins.	$\frac{1}{3}$ of the muffins are banana muffins.

Use the diagram to help you decide whether each student is correct or not.

Tick the box if the student is correct. If the student is not correct, explain what mistake they have made.

Rayshawn ☐ ____________________

Keisha ☐ ____________________

Tareeq ☐ ____________________

Part 4

Making a box for a muffin

Question 5

The muffins have a diameter of 7 cm, and they turn out to be $6\frac{1}{2}$ cm high.

How can you clearly show the height and width on the picture?

a Draw two arrows on the muffin picture. One to show how tall it is and one to show how wide it is.

b The students decide to make small open boxes to hold the muffins. They find this design on the internet.

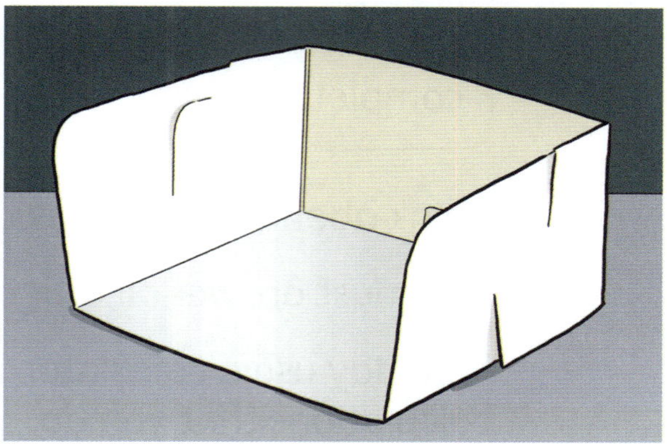

Draw a rough sketch if you need to.

Draw a simple box net (a flat shape that folds into a box). You can draw 5 rectangles: 1 base and 4 sides.

Write the measurements on the net. The box should be big to hold one muffin without wasting card.

Reflection

Give yourself some personal feedback about your work on this task.

Use these sentence stems as starters.

- I am already quite good at ______________________________.
- I can get better at dealing with ______________________________.
- These hints will help me improve ______________________________.

Practice paper 7

This task has 3 parts:

- Part 1 has question 1.
- Part 2 has questions 2 and 3.
- Part 3 has question 4.

→ Before you start ...

Read through the questions.

Find and highlight all the question words (what, which, where, why and when).

Think about how you can answer each question.

Find and circle these instruction words.

match	draw	calculate	give a reason
complete	predict	show	use information

Think about what each word tells you to do.

You must answer all four questions in the spaces provided.

You may use a calculator, but you must write the calculations that you do.

A school weather station

The Grade 4 students are very excited because they are going to set up a weather station at school.

The class will set up their weather instruments in a fenced area next to the school building.

They will measure and record temperature, rainfall, wind speed and wind direction.

You are going to help them read instruments, calculate averages, display data and make decisions about where to place some weather instruments.

Part 1

Setting up a weather station

Question 1

a Match each weather instrument to the correct unit of measurement and what it measures.

Thermometer	Millimetres	How much rain has fallen
Rain gauge	Degrees Celsius	The speed of the wind
Anemometer	Kilometres per hour	How hot or cold it is

Use what you learned in Social Studies to help you here.

b Wind direction tells your where the wind is blowing from. A north wind is a wind that comes from the north.
These arrows show where the wind is blowing to.
Which wind direction is shown by each arrow?

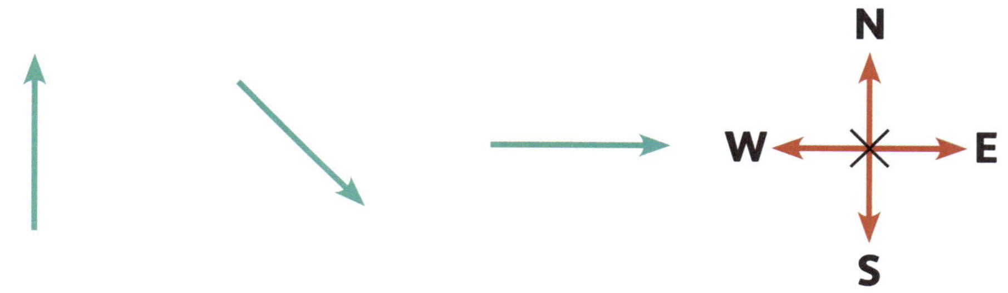

Use the scale on the diagram. You can start by colouring or marking the areas where the rain gauge cannot go to narrow down your choices.

c Here is a plan of the area that the school has set aside for the weather station. The thermometers are placed inside a white wooden box.

The class needs to decide where to place the rain gauge.

- The rain gauge should be placed in an open area with nothing to stop the rain falling into it.
- The rain gauge must be at least 1.5 metres away from any buildings so water does not run off the roof into it.
- The rain gauge must be at least 1 metre away from any trees.
- The gauge should be placed on a pole about 1 metre above the ground to prevent water splashing into it.

Where should they place the rain gauge?

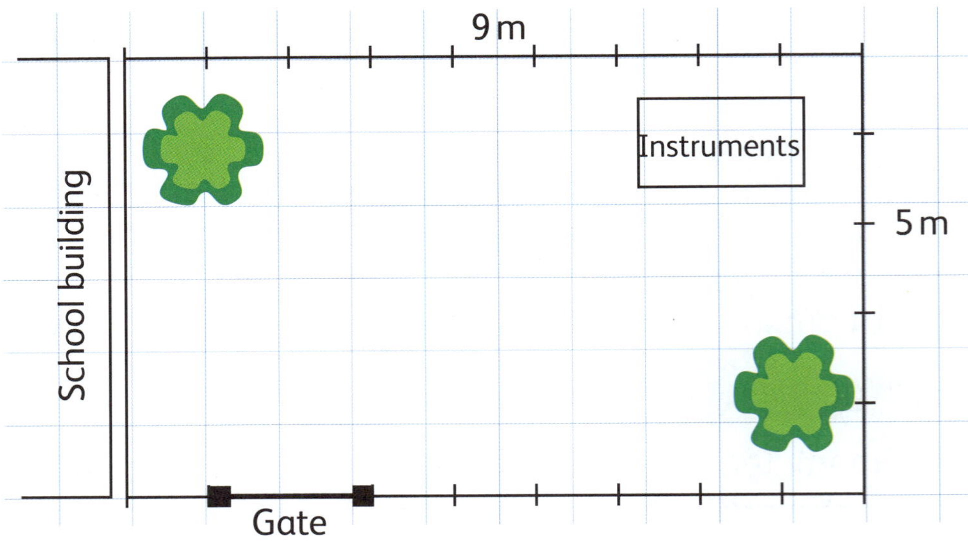

Draw an X on the plan to show where you think they should place the rain gauge.

d The fenced ground set aside for the weather station is 9 metres long and 5 metres wide. What is the area of this space in square metres?

e The school plans to cover half of the weather station area with rubber mats. Each mat covers 1 square metre. How many mats will they need?

Part 2

Working with measurements

Question 2

What does the scale on each gauge show you?

In one week, it rains for three days in a row. The diagram shows how much rain was collected in the rain gauge each day.

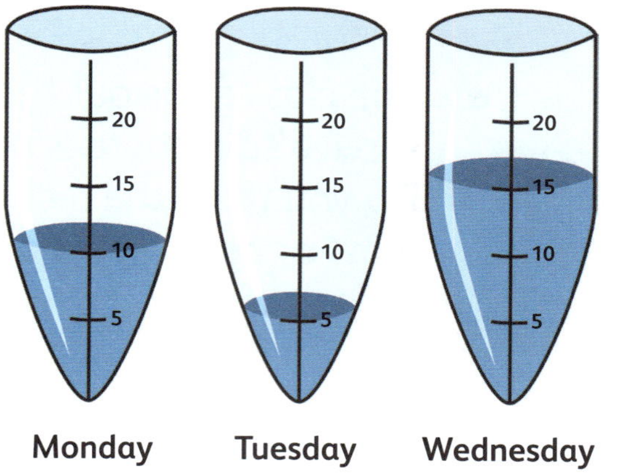

a On which day was 5 mm of rain collected? ______

b What is the total amount of rain collected in the three days?

c Calculate the mean amount of rain for these three days.

d On Thursday it doesn't rain. On Friday they measure 12 mm and over the weekend another 23 mm are collected in the gauge. What is the mean amount of rain per day for that week?

Remember that you can give your own opinion as long as you explain why you think that way.

e The class finds out that in this month the weather service says there is an average of eight days of rain. Do you think the rest of the month is likely to be very rainy or not? Give a reason for your answer.

____

____

You are told he is wrong, so think about what mistake he has made.

f In September, there was 125 mm of rain in Montego Bay. It rained on 12 days that month. Ziggy says, "125 ÷ 12 = 10.4, so it rained 10.4 mm each day." Why is Ziggy wrong?

____

____

Question 3

The class record the temperature each morning and afternoon for four days.

These are their measurements in degrees Celsius (°C).

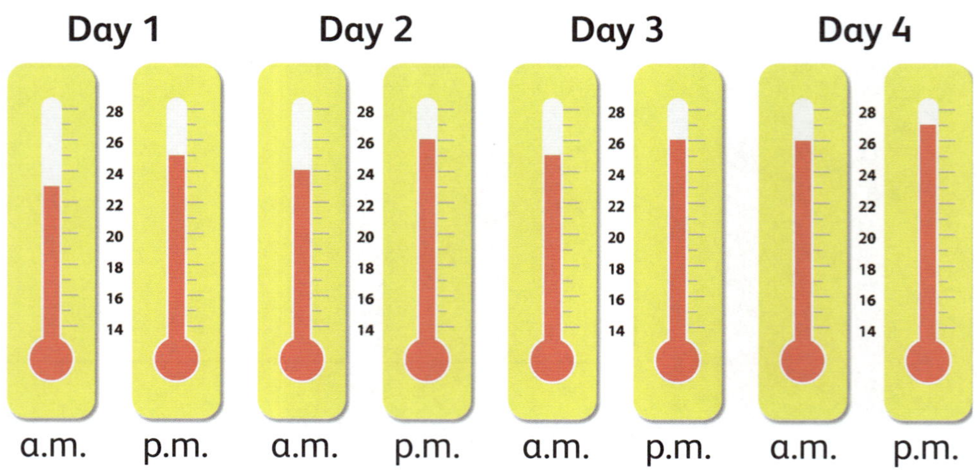

a Michelle says it is usually hotter in the afternoon than in the morning. Is she correct? Explain why or why not.

b What is the median temperature in the morning? Show how you work this out.

Work out the mean for each day before you start drawing your graph.

c Complete this graph to show the mean temperature for each day.

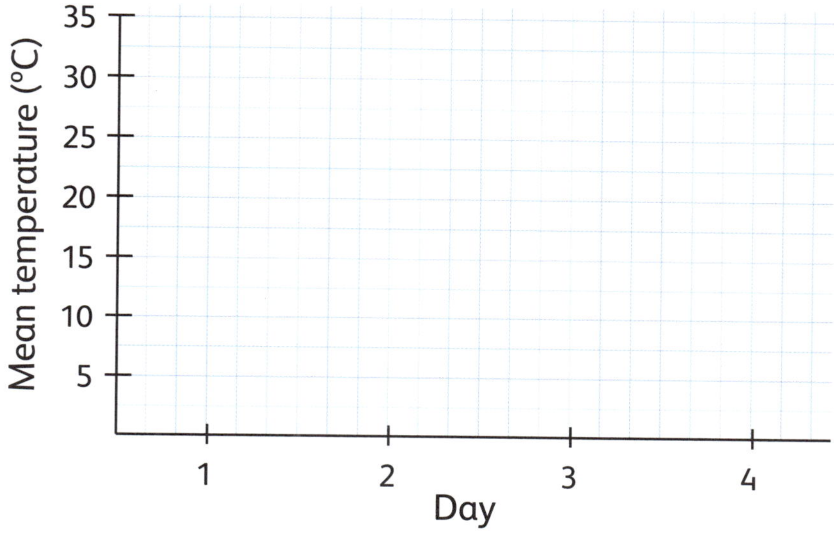

d Predict what the mean temperature will be on day 5. Add this to the graph. Explain why you think it is likely to be this temperature.

How many days are there in March?

e During March, the weather service reports that there is an average of 8 hours of sunlight each day on the South Coast. What is the total number of sunlight hours in March for this region? Show how you work this out.

Part 3

Making sense of weather data

Question 4

Make sure you spend some time reading and making sense of the charts before you try to answer the questions.

Ms Le Croix is Canadian. She wants to spend a week's vacation in Montego Bay, but she is not sure what time of year is best. She wants it to be warm, but not too hot and she is hoping for clear blue skies and that it will not rain much during her vacation.

a Use the information in these two charts to answer the following questions:

i. Which month has the least rainfall?

____

ii. Which months are warmest?

____

iii. Use this to recommend the best time for Mrs Le Croix to book her vacation.

____

Climate chart - Montego Bay

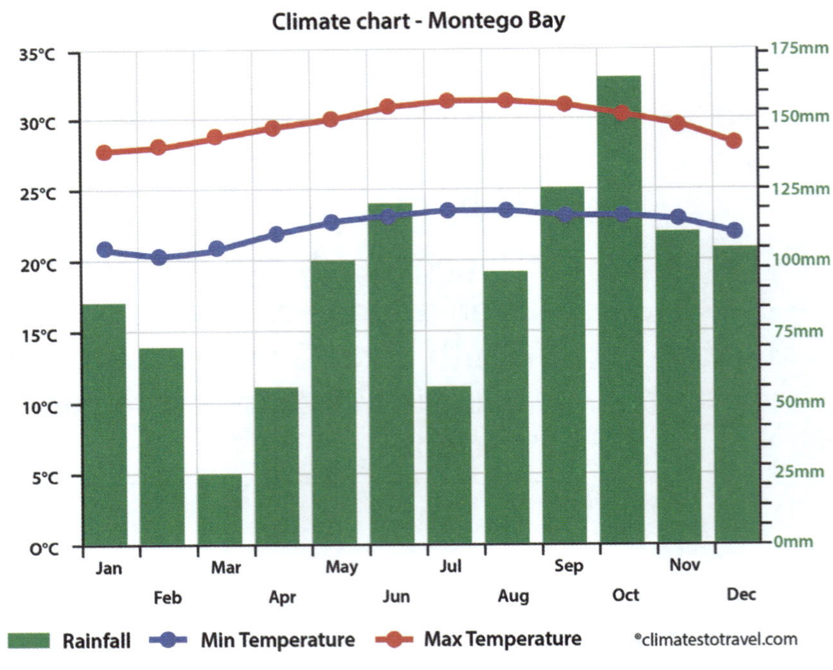

Cloud cover each month of the year

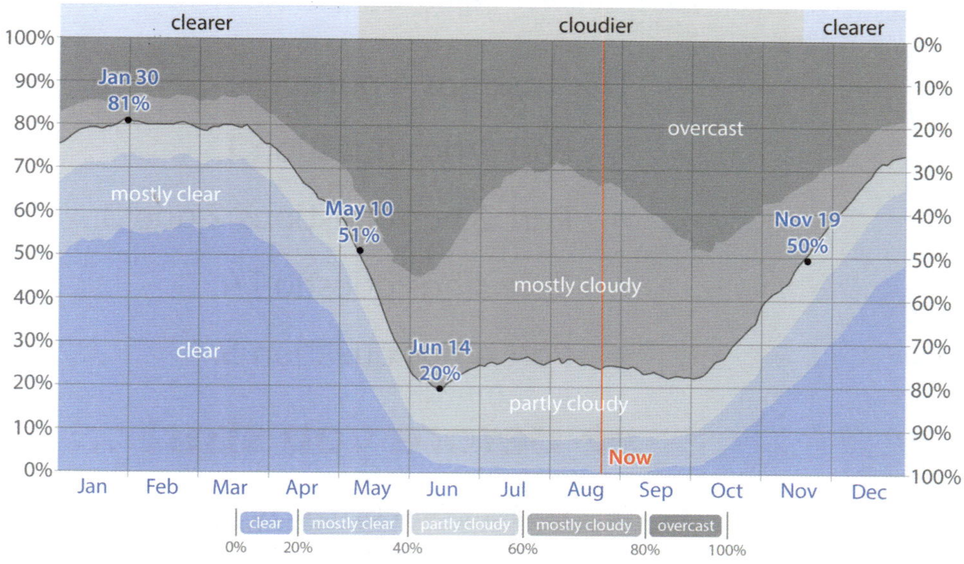

b Mr Jones books a vacation in Montego Bay during May. What weather conditions can he expect?

When is Heroes Day?

c Melinda is interested in history and culture. She wants to visit Jamaica to experience Heroes Day. What weather can she expect at that time of year?

Reflection

You already know that thinking about your mistakes can help you improve.

Find any mistakes in your work.

For each one, decide whether it was:

- **careless** (writing the wrong number, not following instructions, forgetting to give reasons, not showing steps in calculation)
- **calculation error** (miscalculating a value using the basic operations).

Write down three suggestions to help you avoid making these mistakes in the future.

Practice paper 8

This task has 4 parts:

- Part 1 has question 1.
- Part 2 has questions 2 and 3.
- Part 3 has question 4.
- Part 4 has question 5.

➜ Before you start ...

For some questions you may be asked to draw diagrams as part of your answer.

a Label each of these types of diagrams.

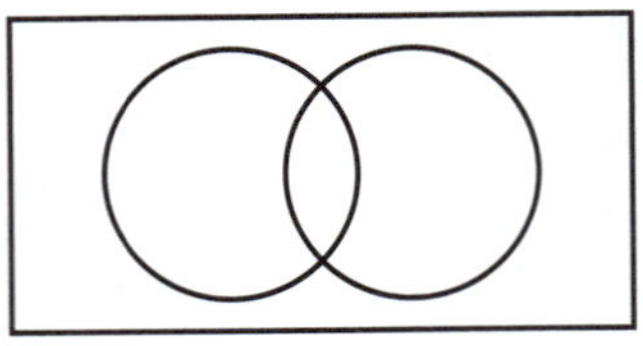

____

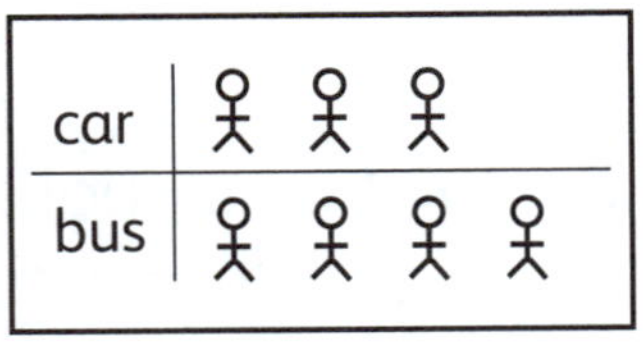

____

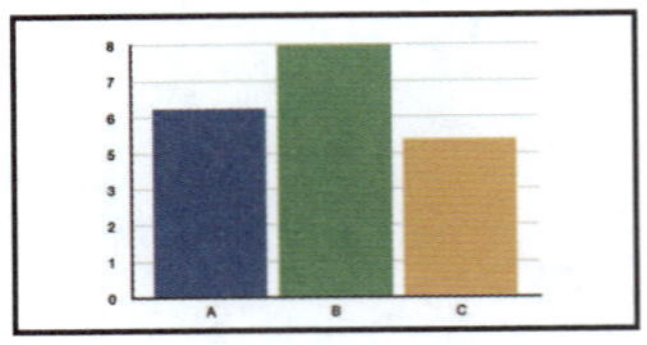

____

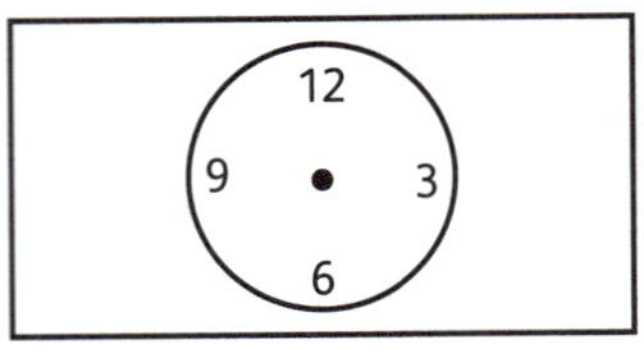

____

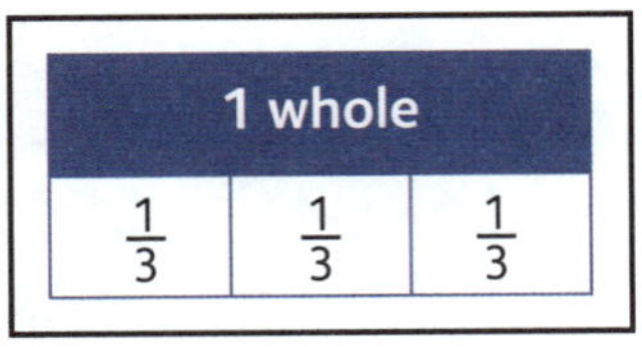

____

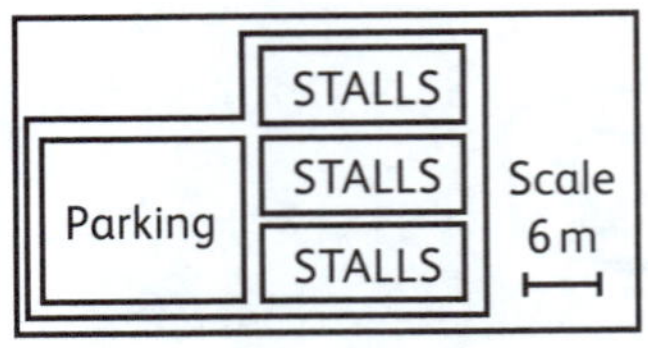

____

b Read through all the questions in this task.

Highlight all the questions where you are asked to draw a diagram.

What type of diagram will you draw for each question? Write your ideas below.

Answer all the questions.

Write or draw your answers in the spaces provided.

Remember to show all your working.

At the market

The students in Class 4 are learning about data handling.

They do a project to find out more about the local market. You are going to help them make sense of the data and answer questions about the market.

Part 1

What is sold at the market?

Question 1

The students did a quick survey to find out what is sold at the local market. This is the data that Jayanne, Samoy and Keenan collected.

Jayanne	Samoy	Keenan
T-shirts	Bananas	Irish potatoes
Carrots	Melons	Sweet potatoes
Apples	Carrots	Apples
Patties	Tomatoes	Carrots
Fruit punch	Sweets	Tomatoes
Cabbage	Plantain chips	Oranges
Yams	Caps	Watermelons
Bananas	Table cloths	Pumpkins
Watermelons	Knitted hats	Peppers
Strawberries		Cucumbers
Straw hats		

Think about what you would do if you were asked to find out what is sold at the local market.

a How do you think they collected this data?

How can you make sure you do not count products more than once?

b How many different products did they find on sale?

You have to think about this and decide what type of graph would be best.

c Design and draw a graph to show how many different fruits, vegetables and other products they found. Label your graph clearly so it is easy to understand.

d Describe what your graph tells you.

Part 2

Who sells what?

Miss Patsy and Miss Flora are both vendors at the market. They sell their goods on Fridays and Saturdays.

Question 2

The picture below shows the types of food items that Ms Patsy and Ms Flora sell.

Think about this. How will you tell which list is for each person?

a Keenan made this table to record what each person sells but he left out some information. Use the picture on the previous page to find the missing information. Write the name of each vendor in the first row. Fill in the missing vegetable in each column.

______	______
______	Yam
Cabbage	Carrot
Sweet potato	Irish potato
Pumpkin	Cabbage
Irish potato	Sweet potato
Tomato	______

Use your general knowledge to answer this.

b Jayanne drew this diagram to show what each person sells.

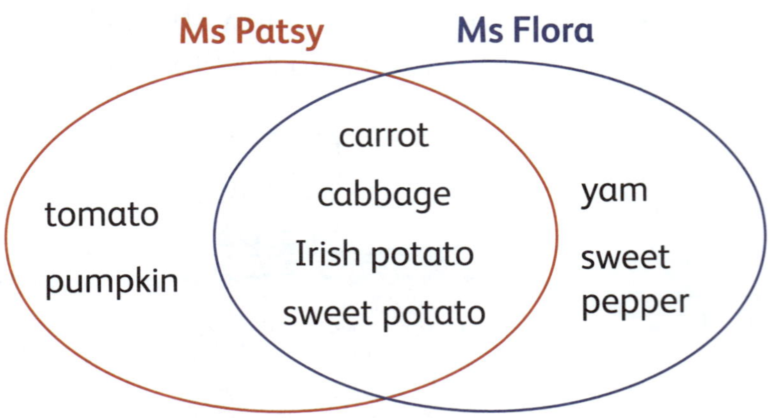

i Why has Jayanne written some of the vegetable in the overlapping part of the circles?

ii Which vegetables does Ms Flora sell that Ms Patsy does not sell? How do you know this?

Use your general knowledge to answer this.

c Ms Campbell wants to buy lychees and guineps but there are none for sale at the market. Suggest one reason why these fruits are not available.

Question 3

a Together, Ms Patsy and Ms Flora take 38 kilograms of carrots to market. Ms Flora took 13 kilograms more than Ms Patsy.
Work out how many kilograms of carrots each person took to the market.

What are you being asked here? Think carefully.

b Mr Raymond wants to buy the following items:

- 3 kg carrots
- $1\frac{1}{2}$ kg cabbage
- $3\frac{1}{2}$ kg sweet potatoes and
- 2 kg Irish potatoes.

These are the prices at each stall.

Ms Patsy	Price per kilogram	Ms Flora	Price per kilogram
Carrot	$300	Yam	$430
Cabbage	$250	Carrot	$280
Sweet potato	$200	Irish potato	$325
Pumpkin	$200	Cabbage	$250
Irish potato	$320	Sweet potato	$170
Tomato	$580	Sweet pepper	$400

To whom would he pay the least amount of money if he bought all the items from her stall?

c Mr Raymond buys from the cheaper stall. He pays for his items with a $5 000 note. What change should he get?

d The vendor had no small bills or coins, so they agreed to round the change to the nearest $100. How much change will Mr Raymond get?

Think about how your family shops. How do they decide who to buy from?.

e Mr Broome decides to buy from Ms Patsy even though some of her prices are slightly higher than Ms Flora's prices. Circle two things that might explain why he does this.

The stall is bigger.

The vegetables look fresher.

She has all the items he wants.

Her produce is not as good.

The goods are packed nicely.

He gets a discount.

Part 3

Getting to market

Question 4

Ms Patsy left her house for the market at 4:37 a.m. and Ms Flora left at 3:55 a.m. Ms Flora arrived at 5:20 a.m. and Ms Patsy took 33 minutes to arrive at the market.

a How long did it take Ms Flora to get to the market?

b Ms Patsy looks at her watch to see what time it is when she arrives at the market. Draw hands on the clock to show what she would see.

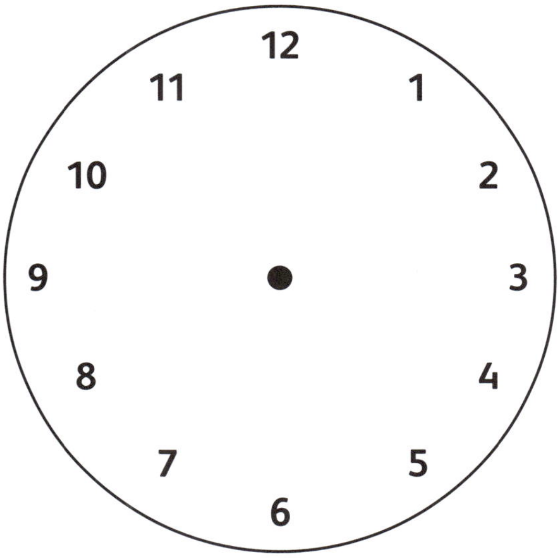

c Which person took the least time to get to the market? Show how you decide.

Show how you decide means the same as show your working out.

50 km per hour means it takes 1 hour (60 minutes) to travel 50 kilometres.

d Ms Patsy travels to the market by bus. The bus travels at a speed of 50 km per hour. How can you use this information to work out how far Ms Patsy travelled from home to the market? Show your ideas here.

Part 4

Growing goods for the market

The students interview some vendors to find out more about how they grow and package the fruits and vegetables they sell.

Question 5

This is a plan of Mr Johnson's plot. He is growing four things at present.

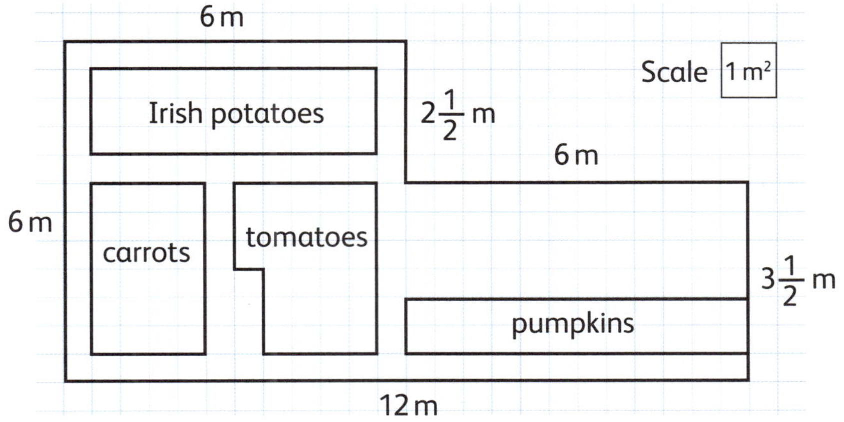

a Which bed takes up the greatest area of the plot? Draw a circle around the plot.

Indicate on the plan means the same as draw your ideas on the plan.

b Mr Johnson plans to plant another bed of pumpkins. He tells the students it is the same area as the existing bed, but that it is not the same shape. Indicate a possible pumpkin bed on the plan.

c Mr Johnson uses $\frac{1}{4}$ bag of manure for every square metre he plants. If he added manure to the existing beds, how much would he need to use?

When you see an incomplete table, it's usually helpful to fill it in.

d Mr Johnson packs tomatoes into a bag. Each bag has a mass of approximately 400 grams.

Number of bags	1	2	3	4	5	6
Mass	400 g					

i What is the mass of 5 bags?

ii How many bags are there if the total mass is 4 800 grams?

iii Mr Johnson picks 10 kilograms of tomatoes. How many 400-gram bags can he fill?

Think about what size bunches people are likely to want. Most people will not want to buy massive bunches of carrots.

e Mr Johnson harvests 168 carrots. He wants to tie them into bunches to sell and he does not want to have loose carrots left over. What size bunches do you think he should make? Why?

Reflection

Kofi says, "the best way to learn something is to do it".

- What do you think he means?
- How can doing practice tasks help you learn more about tasks? Write two ideas.
- How do you learn maths? Write down two things you do.

Practice paper 9

This task has 3 parts:

- Part 1 has question 1.
- Part 2 has questions 2, 3 and 4.
- Part 3 has question 5.

Answer all 5 questions by writing or drawing in the spaces provided.

You will need a ruler.

→ Before you start …

a Complete this checklist before you start the task.

- ☐ I have read the instructions and I have the things I need.
- ☐ I have read through the questions.
- ☐ I have underlined or highlighted important information.

b Look at the questions carefully. Use this table to sort them into three different types of answers.

Simple calculations and answers	Thinking and giving reasons	Many things to work out, more than one possible answer

Arts and crafts

The Grade 4 class is having an arts and crafts day.

The students will take part in two activities: fabric printing and making wooden items.

You are going to help organise the groups, share out the equipment, make sense of instructions and display the work.

Part 1

Underline the instruction words in each question to help you decide how to answer it.

Organising the class

Question 1

There are 32 students in the class. The teacher wants to put them into six groups.

a Explain why the teacher cannot make six groups that each have the same number of students.

____

____

b How would you divide up the class so that each group has approximately the same number of students?

c Each group is going to get the same number of wooden blocks. There are two boxes of blocks, each with 108 blocks. How many should each group get?

d Each group is going to get an equal share of fabric. The fabric is 15 m long. How many metres will each group get?

e Each group is going to get some fabric paint. The paint comes in 1.8-litre containers. What is a fair share of the paint?

Show your working. You will get credit even if your answer is not correct.

f There is green, yellow and black fabric paint.

Each group is allowed to choose two colours. How many different colour combinations are possible?

Part 2

Think about the blocks you cannot see. What will they look like?

Art and craft activities

Question 2

Tameeka's group sticks blocks together to make a larger cube. They put coloured stickers on the outside faces of the larger cube, like this:

a How many small cubes did they use to make the larger cube?

b How many stickers did they need altogether?

c How many cubes will have:

- 1 sticker only?
- 2 stickers?
- 3 stickers?
- 0 stickers?

Use this table to record your answers.

Number of stickers on the cube	1	2	3	0
Number of cubes				

d How can you check that you have the correct numbers? Write your ideas here.

e Nadia makes this smaller cube. She also puts a sticker on each outside face. How will her results be different? Why?

Question 3

Jayson's group decides to make a calendar with the wooden blocks.

They use two blocks to show the date like this:

Each block will have six numbers. One on each face.

Jayson has to work out what numbers to write on the faces of each block so they can make any date from the 1st to the 31st of a month.

He starts like this:

0 1 2 3 4 5 6 7 8 9

once (0) twice (1 2) once (3 4 5 6 7 8 9)

a Explain why Jayson needs the numbers 1 and 2 twice.

____

____

b What numbers should he put on each block? Write them here.

Block 1						
Block 2						

c Does it matter where he writes the numbers on each block? Give a reason for your answer.

____

d The picture shows the same block in three different positions. What number is on the face opposite 3? Explain how you worked this out.

Question 4

Remember symmetrical designs can be folded along a mirror line to get two halves that fit onto each other exactly.

Nicola's group wants to make symmetrical designs on their fabric. They draw these three patterns.

a Draw lines on each pattern to show that it is symmetrical.

Draw lightly in pencil first so you can erase if you make mistakes.

b The students cut out a square of fabric and decide that they will each complete a quarter of the symmetrical pattern. Nicola has completed the first quarter. Help the group by finishing the pattern for them. The purple lines are both lines of symmetry.

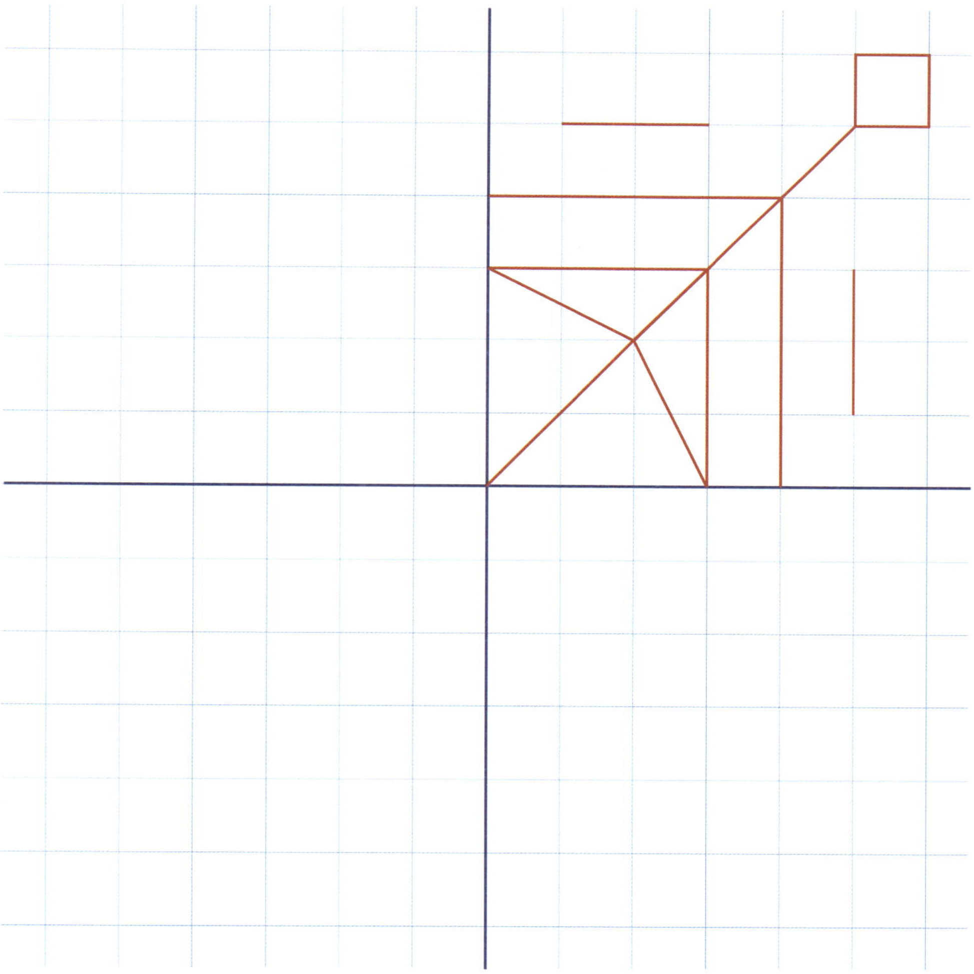

Part 3

Displaying the work

Question 5

a The students decide to display their fabric designs along the passageway at school. They use four pins for one piece of fabric but they overlap the fabrics on the display like this:

The class has 23 pins. How many will they have left after they hang six pieces of fabric?

Can you use a small sketch or model to work this out?

b The wooden objects are displayed on a table. The table is 1.25 m long and 80 cm wide. Each group needs a different amount of space to display what they have made. Read what each group needs and then draw on the diagram to show how you can arrange and fit all six craft projects on the table.

Group A
We need $\frac{1}{10}$ of the area.

Group B
We need a square with sides 30 cm long.

Group C
We need a rectangle 60 cm long and 40 cm wide.

Group D
We need a long narrow area 15 cm wide and 75 cm long.

Group E
We need $\frac{1}{5}$ of the area.

Group F
Our area is 20 cm wide and 40 cm long.

There is more than one way to do this. Plan carefully before you draw your final answer.

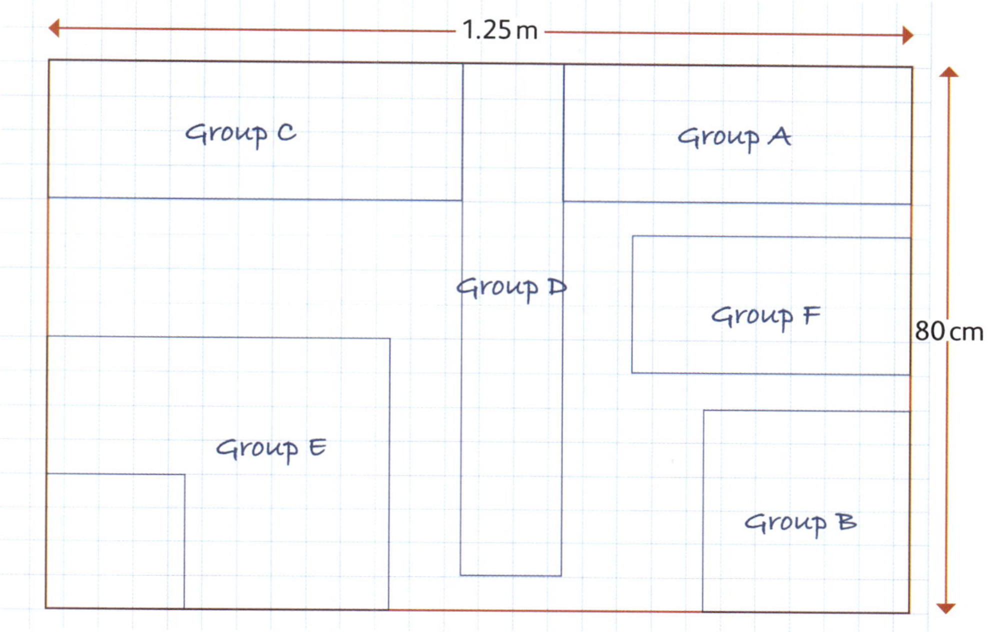

c How can you work out the area of the table that is not covered by the displays? Suggest two different ways to do this.

____

____

Reflection

- Write down three things in this task that you did well.

____

- What could you do better?

____

- Write down one thing that you learned in another task that helped you here.

____

Practice paper 10

This task has 3 parts:

- Part 1 has questions 1 and 2.
- Part 2 has questions 3 and 4.
- Part 3 has question 5.

You must answer all of the questions.

Write or draw your answers in the space provided.

Remember to always show your working.

→ Before you start …

You have learned a lot about doing tasks in this book.

Write down three important things that you should always do before you start a task.

1 ____________________

2 ____________________

3 ____________________

Why is it important to do these things?

____

A new play area

The local community are planning a new play area in an existing park for the local children.

They will supply some equipment, but they have asked the community to help them design and plan the layout of the play area.

Part 1

Planning and mapping the play area

Question 1

The council will supply a swing set, a roundabout, a slide and a climbing gym.

The amount of space that each piece of equipment will take up is shown on the grid below.

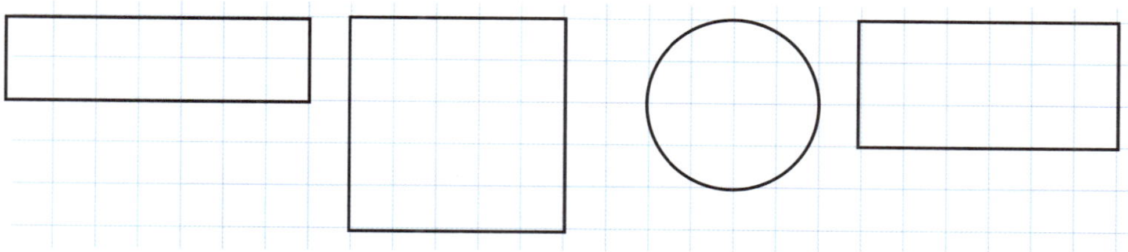

Use the squares on the grid to work out the area each piece of equipment needs.

a Write the name of each piece of equipment below the space it will need.

An area of the park has been covered with a soft rubber material for safety. A plan of the area is shown below.

The equipment must all go into this area. There needs to be some space around each piece of equipment for safety reasons.

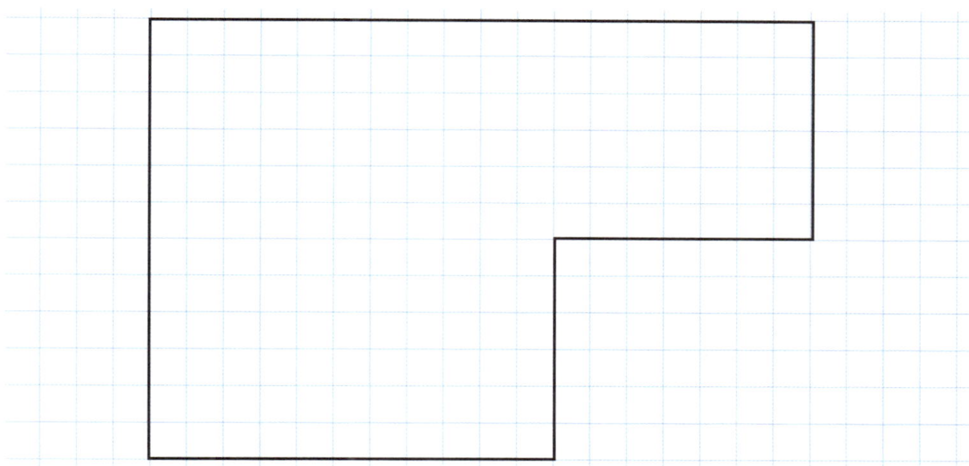

b Where should each piece of equipment go? Draw your suggestion on the plan on the previous page.

c Give two reasons why you arranged the equipment like this.

Reason 1

___

Reason 2

___

Question 2

The council also wants to change a paved area near the playground by adding a flower bed. They show the community this plan of the paved area and the flower bed.

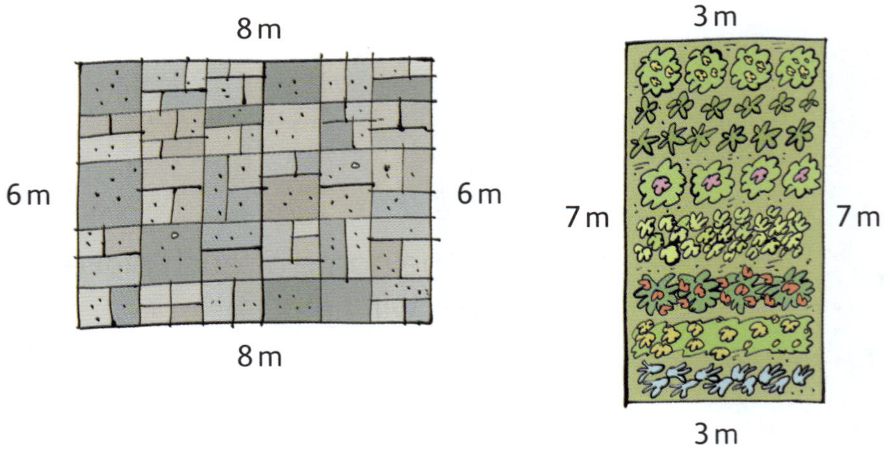

a Is it possible to fit this flower bed into the paved area? Explain your answer and draw a sketch to show why the flower bed will or will not fit.

Show your working clearly.

b Ms Williams wants to know how big the whole park is. She takes two large paces and measures how far she has travelled.

Ms Williams paces out the perimeter of the park. The park is shaped like a rectangle and it is 40 paces wide and 100 paces long.

Work out the perimeter and area of the park.

c Sandra also paces out the perimeter of the park. She says Ms Williams' measurements must be wrong because the park is 48 paces wide and 124 paces long. What mistake has Sandra made in her thinking?

Part 2

When you are given mixed fractions, you need to think about equivalent fractions.

Playing in the park

Question 3

The children's picnic area is going to be tiled. The council provides boxes of tiles in different shapes and colours.

The first box of tiles is $\frac{2}{5}$ square, $\frac{2}{5}$ rectangular and $\frac{1}{5}$ triangular. $\frac{1}{4}$ of the tiles are black, $\frac{1}{4}$ are green and $\frac{1}{2}$ are yellow.

Draw the smallest set of tiles that could be in the box.

Question 4

Josh and Kofi measure the length of the shadow made by a metre high fence post and find that the shadow is 50 cm long.

a At the same time, the climbing gym makes a shadow that is 1.25 m long. How tall is the climbing gym?

b Sanita measures the shadow of a hibiscus bush and finds that it is 68 cm long. She also measures the shadow of a palm tree and finds that it is 1.35 m long. Sanita says this means the palm tree is twice as tall as the hibiscus bush. Is she correct? Give a reason for your answer.

Part 3

Finding out what people want

Class 4 think that it is important to find out what changes and features people want to see in the park. They decide to do an investigation to find out.

Question 5

a Read what three of the students say:

i What is a survey?

____

ii What is a sample?

____

iii Do you agree that the students should focus on people who use the park or who live near it? Give a reason for your answer.

____

____

Always read the labels and the scale when you work with graphs.

b The students collect data over the weekend. They keep a tally of their results and then draw these graphs to show the data.

What adults want

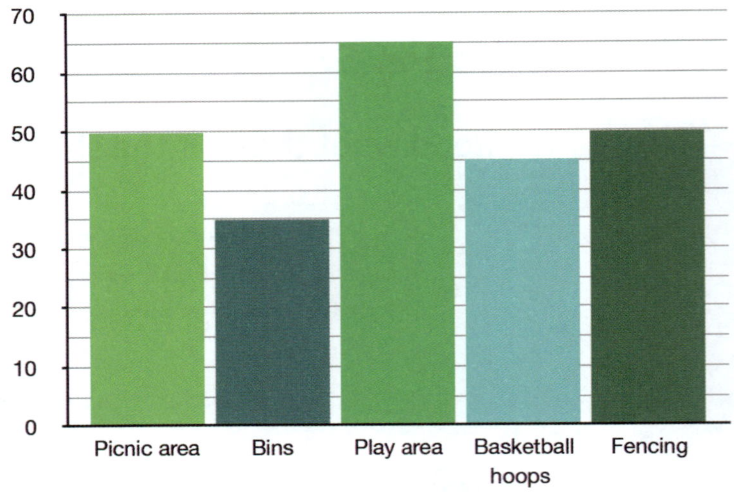

What children want

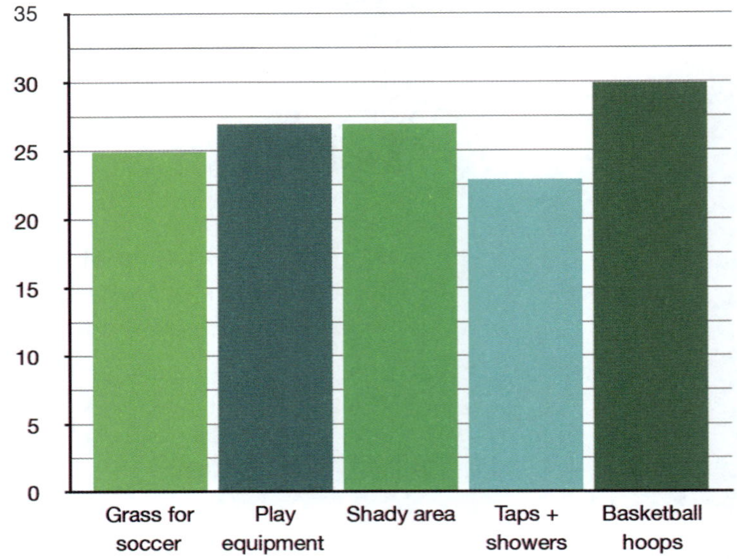

Answer these questions using information from the graphs.

i What feature do most adults want? ____________

ii How many adults think fencing the park is important? ____________

iii What is more important to adults: bins or a picnic area? Give a reason for your answer.

____

iv How many children want play equipment in the park? ____________

v How many people altogether want basketball hoops? ______

c Maria says that adults and children want different things. Nina agrees, but she says they also want some of the same things. Choose three things that the council could provide in the park that you think would make adults and children happy.

______

d Once they have done the survey, the students decide that they need to ask some more questions to help them decide what people want most. Write down two questions they could ask to find this data.

______

______

e Look at this diagram showing the steps in a statistical investigation. Match what the students did to each step in the data cycle. Write your answers in the spaces on the diagram.

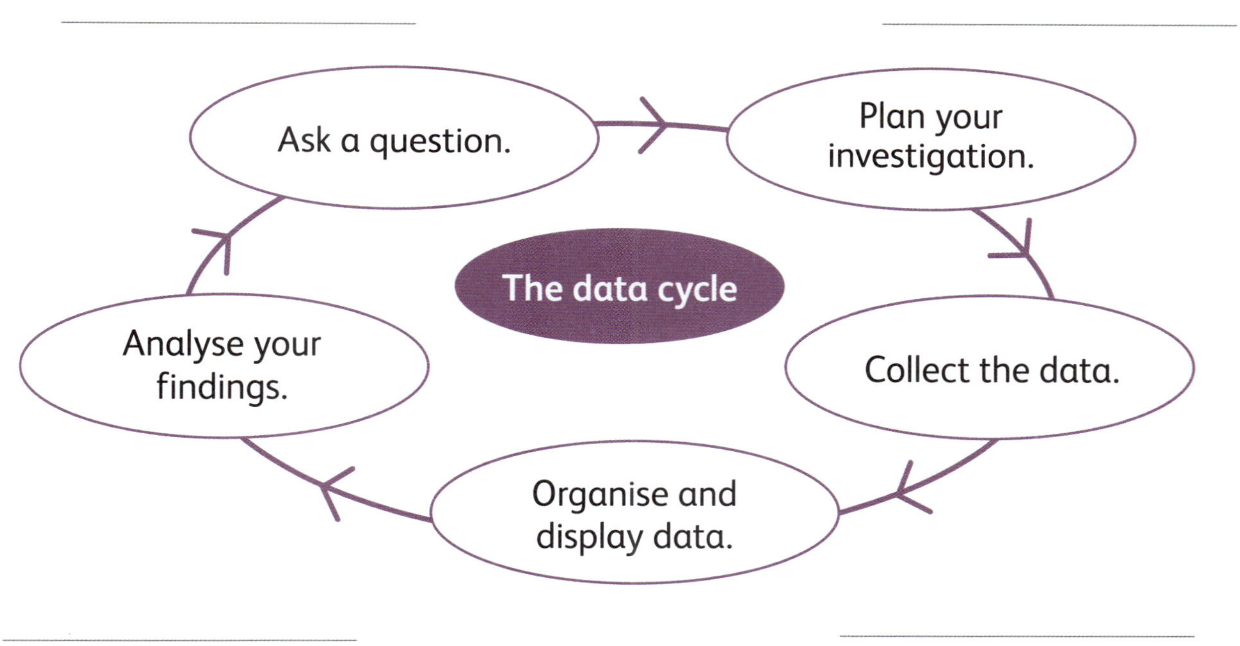

Reflection

- What did you find most difficult or confusing in this task?

  ____

- What did you like about it?

  ____

- How did this help you?

  ____